Workshop on Micro-Mini and Nano Dosimetry (MMND 2018)

AA006641

Journal of Physics: Conference Series
Volume 1154

Mooloolaba, Australia
6 – 11 February 2018

ISBN: 978-1-5108-8335-2
ISSN: 1742-6588

Printed from e-media with permission by:

Curran Associates, Inc.
57 Morehouse Lane
Red Hook, NY 12571

proceedings.com

Some format issues inherent in the e-media version may also appear in this print version.

Printed by Curran Associates, Inc. (2019)

For permission requests, please contact the Institute of Physics at the address below.

Institute of Physics
Dirac House, Temple Back
Bristol BS1 6BE UK

Phone: 44 1 17 929 7481
Fax: 44 1 17 920 0979

techtracking@iop.org

Additional copies of this publication are available from:

Curran Associates, Inc.
57 Morehouse Lane
Red Hook, NY 12571 USA
Phone: 845-758-0400
Fax: 845-758-2633
Email: curran@proceedings.com
Web: www.proceedings.com

Table of contents

Volume 1154

Micro-Mini & Nano Dosimetry (MMND) 2018

6–11 February 2018, Mooloolaba, Queensland, Australia

Accepted papers received: 3 December 2018
Published online: 7 March 2019

Preface

Papers

IOP Conf. Series: Journal of Physics: Conf. Series **1154** (2019) 011001 doi:10.1088/1742-6596/1154/1/011001

Introduction to Mini-Micro & Nano Dosimetry and Innovative Technologies in Radiation Oncology (MMNDITRO) 2018

The biannual Micro-Mini & Nano Dosimetry and Innovative Technologies in Radiation Oncology (MMNDITRO2018) international workshops, traditionally organized by the Centre for Medical Radiation Physics (CMRP), University of Wollongong jointly with the Memorial Sloan Kettering Cancer Center (MSKCC), New York were held from 6-11 February 2018 at the Mantra Mooloolaba Beach Resort, Mooloolaba, Queensland, Australia.

Continuing our series of biennial meetings, the MMND 2018 workshop, from 6-8 February 2018, brought together both Australian and international radiation oncologists, medical physicists, radiation scientists and nanomedicine experts to discuss advances in radiation oncology modalities and radiation dosimetry technologies for quality assurance in radiation therapy, radiobiological optimisation of treatments, and other relevant technologies, to further improve clinical outcomes of radiation therapy. For the first time in the MMND workshop series, a special brachytherapy day, led by Dr Michael Zelefsky MD, was held.

The ITRO 2018 workshop, from 10-11 February, was devoted to clinical implementation of new technologies in X-ray therapy, brachytherapy and particle therapy. A special focus of ITRO was Integrating biology and technology including immunotherapy, SBRT, hypofractionation, particle therapy, SRS, Integration of multi-parametric MRI into radiotherapy and other approaches on the way to personalised treatments.

For more information about MMNDITRO2018 program visit https://mmnditro2018.dryfta.com/en/

Anatoly Rozenfeld PhD
Dr Joshija (Josh) Yamada MD

Co-Chairs of MMND ITRO 2018 Dr Joshija (Josh) Yamada MD (MSKCC) and Distinguished Professor Anatoly Rozenfeld PhD (CMRP).

Dr Michael Zelefsky MD

Peer review statement

All papers published in this volume of *Journal of Physics: Conference Series* have been peer reviewed through processes administered by the proceedings Editors. Reviews were conducted by expert referees to the professional and scientific standards expected of a proceedings journal published by IOP Publishing.

IOP Conf. Series: Journal of Physics: Conf. Series **1154** (2019) 012001 doi:10.1088/1742-6596/1154/1/012001

Fibre optic dosimetry in synchrotron microbeam radiation therapy

J Archer[1], E Li[1], A B Rosenfeld[1,2] and M Lerch[1,2]
[1]Centre for Medical Radiation Physics, University of Wollongong, NSW 2522, Australia
[2]Illawarra Health and Medical Research Institute, University of Wollongong, NSW 2522, Australia

E-mail: enbang@uow.edu.au

Abstract. Synchrotron microbeam radiation therapy is a novel pre-clinical therapy method that uses high brilliance, spatially fractionated, low energy x-rays to deliver a very high dose rate. A conventional spatial fractionation for these beams is 50 μm microbeam width at 400 μm peak-to-peak separation. To perform dosimetry on these beams, a dosimeter with a high spatial resolution is required. We present a plastic scintillator fibre optic dosimeter that has been demonstrated to be able to resolve the microbeam structure. The advantages of such a dosimeter is the water equivalence, passive components exposed to the radiation field, relatively inexpensive components and simple fabrication. We use cylindrical BC-400 plastic scintillator optically coupled to a 1 mm diameter core plastic optical fibre. BC-620 reflective paint was coated on the end of the probe to maximise the capture of light. A silicon photomultiplier (SensL MiniSM) was used to measure the light output of the probe. A 50 μm resolution probe and 20 μm probe was tested at the Australian Synchrotron, on the Imaging and Medical Beam-Line. We were able to resolve microbeams with both resolution probes. The depth dose measurements matched that of a Pinpoint ionisation chamber except for an over-response at shallow depth.

1. Introduction

Synchrotron microbeam radiation therapy (MRT) is a novel pre-clinical therapy method that uses high brilliance, spatially fractionated, low energy synchrotron x-rays to deliver a very high dose rate within the microbeams. A conventional spatial fractionation for these beams is 50 μm microbeam width at 400 μm peak-to-peak separation (which is used on the Imaging and Medical BeamLine at the Australian Synchrotron). To perform dosimetry on these beams, a dosimeter with a high spatial resolution and radiation tolerance is required. We present a plastic scintillator fibre optic dosimeter that has been demonstrated to be able to resolve the microbeam structure. The advantages of this dosimeter is its water equivalence, passive components exposed to the radiation field, relatively inexpensive components and simple fabrication.

Plastic scintillator dosimeters have been researched on for over two decades [1-3]. They have been applied to a range of medical radiation therapy fields, such as external beam therapy [4-6], brachytherapy [7, 8] and prostate cancer therapy [9]. The challenge to applying this type of dosimeter to MRT is the small sensitive volume required to properly resolve microbeams, which reduces the overall light signal that can be measured. We present the highest spatial resolution plastic scintillator fibre optic dosimeter found in the literature to date.

IOP Conf. Series: Journal of Physics: Conf. Series **1154** (2019) 012001 doi:10.1088/1742-6596/1154/1/012001

2. Materials and Methods

We use cylindrical BC-400 plastic scintillator optically coupled to a 1 mm diameter core Eska CK-40 plastic optical fibre (illustrated in figure 1). BC-620 reflective paint was coated on the end of the probe to maximise the capture of light. A silicon photomultiplier (SensL MiniSM) was used to measure the light output of the probe. The use of plastic components ensures that the probe is water equivalent for radiation interaction. BC-400 is energy independent over a large range of energies, linear, and temperature independent over 0-60 °C [1]. The maximum one dimensional spatial resolution of the probe is determined by the thickness of scintillator used. We present the results from two probes, with 50 μm and 20 μm spatial resolutions, with measurements of the intrinsic microbeam profile, and depth dose curves.

The intrinsic profile was measured by simply scanning the probe laterally across the field and sampling the detector response at 1000 Hz. The time scale is converted to position using the speed of the scanning motor. The depth dose curves were obtained by scanning the detector vertically through the field at each depth and integrating the response.

To reduce noise in the intrinsic microbeam profiles, a smoothing filter is run across the raw data. We use an unweighted moving average with a window size of 17 was used (further details on this method can be found in Archer et al 2017 [5]).

The experiments were performed at the Australian Synchrotron, on the Imaging and Medical BeamLine. The 50 μm probe was tested using Gammex Solid Water to provide attenuation and backscatter, while the 20 μm probe was tested in a water tank. To test the effect of the BC-620 paint on the light capture, the depth dose measurements with the 20 μm probe were repeated with and without the paint.

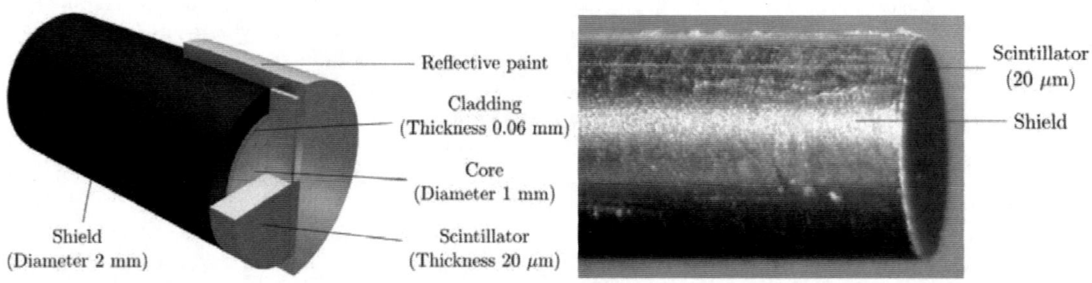

Figure 1. Left: Polished scintillator coupled to large core fibre (not to scale). Right: Microscope image of the probe tip, without the reflective paint. The scintillator is the dark, thin layer on the end. For maximum one-dimension resolution, the radiation beam must be incident perpendicular to the fibre axis. Reproduced from Archer et al [10].

3. Results and Discussion

We were able to resolve microbeams with both resolution probes (shown for the 50 μm and 20 μm probes in figure 2). The average microbeam widths measured with the 50 μm probe was (63 ± 2) μm, which compares favourably with a silicon strip detector of the same resolution [11, 12], which measured (62.4 ± 0.9) μm. The 20 μm probe measured the widths to be (52.1 ± 6.5) μm. The depth dose measurements with the 50 μm probe matched that of a Pinpoint ionisation chamber except for an over-response at lower depth. This result was similar to the 20 μm probe in water. The 20 μm probe was also scanned continuously through depth (Figure 4), which resulted in a similar over-response. Removing the BC-620 paint has no significant effect on the depth dose shape but reduced the total light output by 24% (seen in Figure 3).

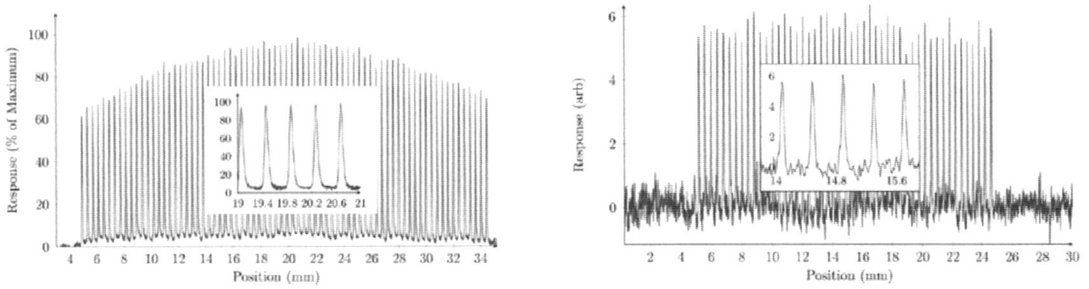

(a) (b)

Figure 2. (a) The intrinsic microbeam profile measured with the 50 μm resolution probe [13]. (b) The profile with the 20 μm probe [10]. Insets show the centre section of the microbeam scans.

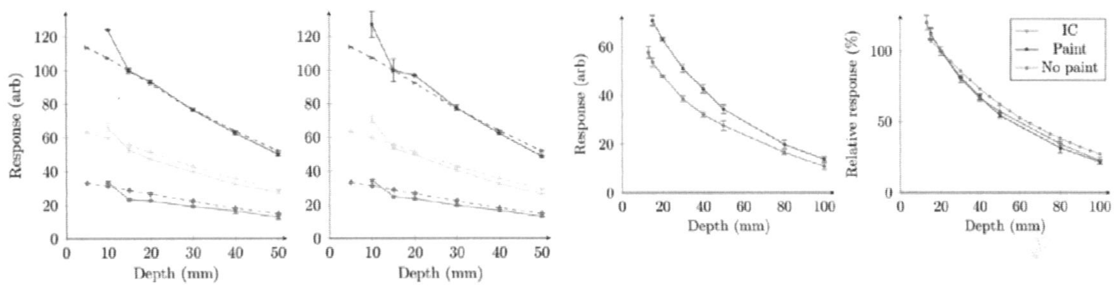

(a) (b) (c) (d)

Figure 3. Depth dose responses collected by the 50 μm probe scanning at (a) 10 mm/s and (b) 5 mm/s [13]. (c) 20 μm probe relative response with and without reflective paint. (d) Same probe normalised to 20 mm depth against ionization chamber [10]. Three BDAs are used with the 50 μm probe: 2.014 mm, 1.052 mm and 0.532 mm.

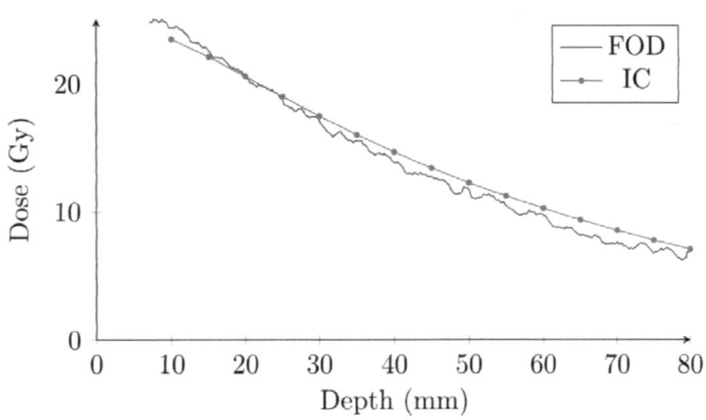

Figure 4. Continuous scan of the 20 μm probe in face-on mode, from 7 mm to 80 mm depth. Also shown is results from ionisation chamber. The 20 μm probe is calibrated to 20 mm depth against the ionisation chamber.

4. Conclusions

The results presented here demonstrates the ability for the fibre optic dosimeter probe to be able to effectively resolve microbeams. The depth dose results show a consistent discrepancy to the ionisation chamber results at low depths. The comparison with and without the BC-620 paint demonstrates that any radiation hardening effects cannot explain this discrepancy. The higher sensitive volume of the ionisation chamber makes a direct comparison of the results challenging.

The reduction in light output by removing the BC-620 paint has two reasons; a lower amount of light being reflected back into the fibre, and the higher Z material of the paint (TiO_2) having a higher dose enhancement. As one of the main advantages of using a probe of this design is the water equivalence, not using the BC-620 paint is justified to remove any possible unwanted dose enhancements.

We anticipate that this dosimeter design can be refined to the point that it can be used in clinical quality assurance in the challenging field of MRT dosimetry. This will provide a relatively inexpensive, simple dosimetry tool to medical physicists responsible for MRT treatments.

5. Acknowledgements

This project was supported by UOW's Global Challenges Program. This research was undertaken on the Imaging and Medical BeamLine at the Australian Synchrotron, Victoria, Australia (AS162/IMBL/10829,12154). Authors JA, EL, and ML acknowledge the support of the Australian NH&MRC (APP1093256). This research has been conducted with the support of the Australian Government Research Training Program Scholarship.

6. References

[1] Beddar A S *et al* 1992 *Phys. Med. Biol.* **37** 1883-1900
[2] Beddar A S *et al* 1992 *Phys. Med. Biol.* **37** 1901-13
[3] Beddar A S 2006 *Radiat. Meas.* **41** S124–33
[4] Archer J *et al* 2017 *Med. Phys.* **44** 1965-8
[5] Archer J *et al* 2017 *Phys. Medica* **42** 185-8
[6] Kim K A *et al* 2013 *Radiat. Prot. Dosim.* **153** 294-9
[7] Sliski A *et al* 2005 *Med. Phys.* **32** 2137
[8] Hurley C *et al* 2006 *Nucl. Instrum. Meth. A* **565** 801-11
[9] Klawikowski S J *et al* 2014 *Phys. Med. Biol.* **59** N27-36
[10] Archer J *et al* 2018 *J. Synchrotron Radiat.* **25** 826-832
[11] Fournier P *et al* 2017 *Radiat. Meas.* **106** 405-11
[12] Lerch M L F *et al* 2017 *J. Phys.: Conf. Ser.* **777** 012009
[13] Archer J *et al* 2017 *Scientific Reports* **7** 12450

Today's monolithic silicon array detector for small field dosimetry: the Octa

G Biasi[1], K Al Shukaili[1,2], M Petasecca[1], S Corde[1,3], S Guatelli[1], V L Perevertaylo[4], T Kron[1,5,6] and A B Rosenfeld[1]
[1]Centre for Medical Radiation Physics, University of Wollongong, NSW, Australia
[2]National Oncology Centre, Royal Hospital, Muscat, Oman
[3]Department of Radiation Oncology, Prince of Wales Hospital, NSW, Australia
[4]SPA-BIT, Kiev, Ukraine
[5]Peter MacCallum Cancer Centre, Melbourne, VIC, Australia
[6]Sir Peter MacCallum Cancer Institute, University of Melbourne, VIC, Australia

E-mail: gb446@uowmail.edu.au

Abstract. The dosimetry of small photon beams is challenging due to detector position uncertainties, dose averaging and lack of electronic equilibrium. Currently only few, single detectors are suitable for measurements in this context, and none is ideal. This study reports on the dosimetric characterization of small fields collimated by fixed cones, performed by a novel 2D monolithic silicon array detector, the Octa.

1. Introduction

The challenges associated with the relative dosimetry of small photon beams, i.e. detector position uncertainties, dose averaging and lack of charged particle equilibrium, have been widely discussed in the recent literature [1,2]. For a detector dedicated to small field dosimetry, characteristics such as a sensitive volume (SV) sufficiently small with respect to the radiation field and the ability to offer high spatial resolution measurements are considered paramount. Ideally, it would also be water equivalent and have a response which is linear with the absorbed dose, as well as be energy and dose rate independent [3,4]]. However, no ideal detector dedicated to small field dosimetry exists, and it is advised to use different dosimeters and cross-check the consistency of results [1,2].

In particular, solid-state detectors are recommended by the IAEA-AAPM protocol dedicated to small field dosimetry [2], but only single 1D solid-state detectors used with various scanning techniques have been shown to offer the necessary sub-mm spatial resolution [1]. Furthermore, their readings need correction factors to account for beam perturbations that are detector design, linear accelerator (linac) treatment head design, beam quality, field size and measurement conditions dependent [5]. As a consequence, these are inconvenient to use in practice because of the multidimensional factor dependencies (field size, depth and distance) [6].

A 'correction-free' detector, i.e. one maintaining a correction factor close to unity, would be a preferable solution. This has been shown to be possible with the addition of low density media to the high density SVs components [7]. However, it would still be necessary to verify that these modifications are correctly compensating for a specific measurement condition [8].

Recently, the Centre for Medical Radiation Physics (CMRP) has developed the Octa, a 2nd generation silicon array detector dedicated to small field dosimetry which has been shown to be accurate for relative dosimetry with both flattened and flattening filter free photon beams [9] and to possess unique potentials for quality assurance for an Accuray CyberKnife system [10]. This study reports on its dosimetric characterization of small fields collimated by fixed cones attached to an Elekta Axesse linac.

2. Material and Methods

The Octa, pictured in Figure 1, is a 2D monolithic silicon array detector based on 512 diode-SVs arranged on 4 arrays. The device offers sub-mm spatial resolution, with diodes having a 0.3 mm pitch along the vertical and horizontal arrays and a 0.43 mm pitch along the 2 diagonals. It is operated in passive mode, connected to a multichannel data acquisition system based on a commercially available analogue front end (AFE0064, Texas Instruments). Conceived for measurements in solid water, the device is sandwiched between two Perspex plates each 5 mm thick and has a small air gap on top of its SVs to minimize the corrections that are required to relate its readings to dose [9].

Experimental measurements described in this study were carried out at the Prince of Wales Hospital, Randwick (NSW), using 6 MV flattened photon beam from an Elekta Axesse linac with a retrofitted Agility head. Fixed conical collimators (Elekta) with nominal diameter, defined as the projection of their openings at the isocentre, between 5 mm and 50 mm were employed. Parameters commonly used by commercial treatment planning systems, such as dose profiles (DPs), percentage depth dose (PDD) and output factors (OFs) were measured with at least 2 different detectors. For the Octa, measurements were performed in solid water (*Gammex RMI 457, Middleton, USA*). Measurements by GafChromic EBT3 films (ASHLAND) performed in solid water (*Gammex RMI 457, Middleton, USA*) and by a SFD diode (IBA Dosimetry) performed in water tank (*Bluephantom, IBA*) were added to the study to cross-check the consistency of results. EBT3 GafChromic films were scanned 24 hours post irradiation with an EPSON 10000XL using a 48-bit RGB with a resolution of 72 dpi following a procedure detailed in [11]. A test to verify the gantry sag was performed by rotating the gantry at 0° and 180° and measuring the shift in the most responsive SV on the Octa.

Prior to all measurements, the Octa was aligned with respect to the machine central axis (CAX) by maximizing the response of its central pixel using the smallest available field size. OFs and DPs were measured at 90 cm source to surface distance (SSD), 10 cm depth. The accuracy of DPs was assessed by comparing FWHM and penumbra widths, defined as the distance between the 20% and 80% of the CAX dose. CAX PDD were measured with 10 cm solid water for backscattering purposes and by adding the required amount of solid water slabs on top of the detector. For a quantitative estimation of the results, the detector readings were analysed using MATLAB (Mathworks) with a shape preserving interpolant function.

3. Results

The central pixels of the Octa were small enough to identify the CAX position accurately without any volume-averaging effect. Figure 2 shows OFs measured by the Octa and EBT3 films for all fixed cones investigated, normalized to the biggest available field size of 50 mm diameter. Figure 3 shows DPs measured by the Octa and EBT3 films for a 5 mm diameter fixed cone. In the figure, profiles are normalized to the response at CAX and aligned such that the origin lies at the coordinate corresponding to the 50% CAX dose. Figure 4 shows PDD measured by the Octa and a SFD diode for a 5 and a 10 mm diameter fixed cone. Nominal depths were converted to water equivalent depths to account for the density of the Perspex plates of the Octa.

IOP Conf. Series: Journal of Physics: Conf. Series **1154** (2019) 012002 doi:10.1088/1742-6596/1154/1/012002

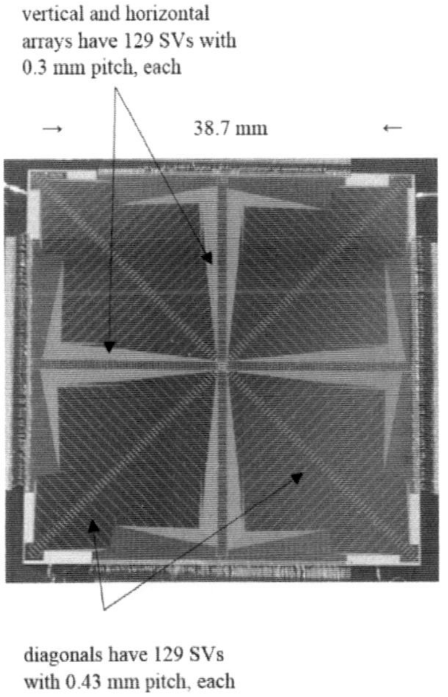

Figure 1. The Octa is a 2D monolithic silicon array detector consisting of 512 diode-SVs arranged along 4 intersecting orthogonal linear arrays.

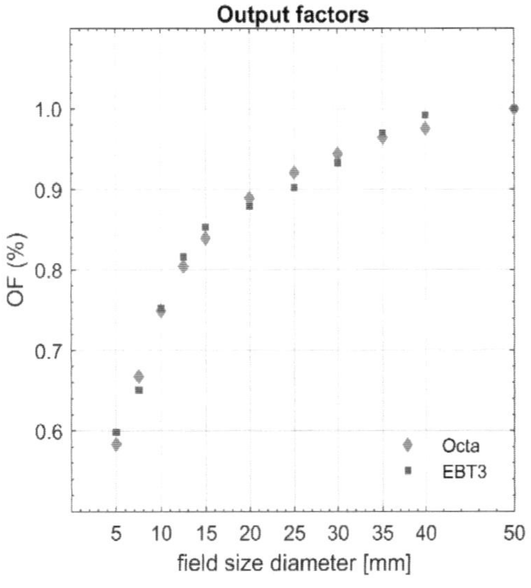

Figure 2. OFs measured by the Octa and EBT3 films for circular fields from 5 mm to 50 mm diameter, collimated by fixed conical cones. OFs are shown normalized to the 50 mm diameter response.

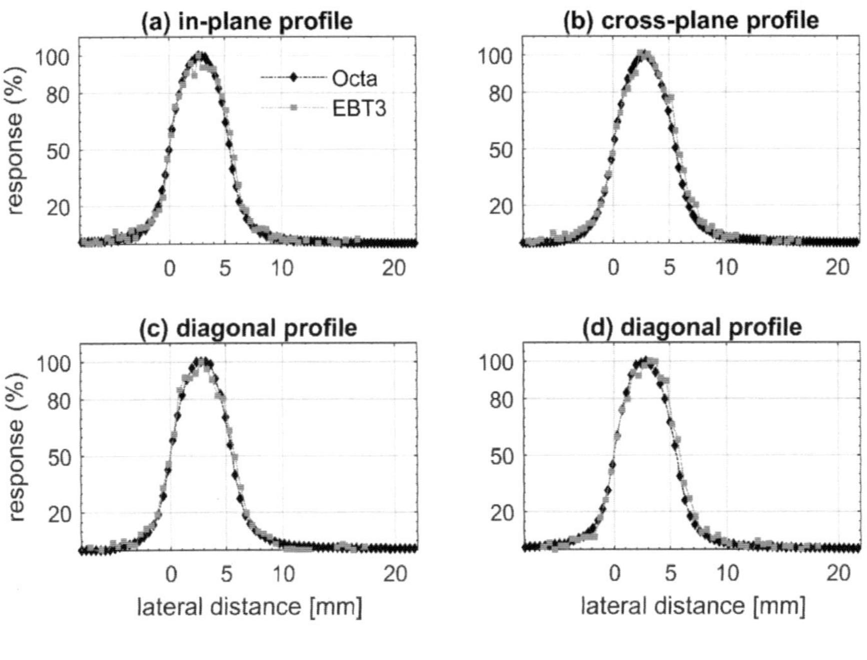

Figure 3. DPs measured by the Octa and EBT3 films for a 5 mm diameter circular field collimated by a fixed conical cone. Profiles are normalized to the CAX dose response and aligned to its 50% value.

IOP Conf. Series: Journal of Physics: Conf. Series **1154** (2019) 012002 doi:10.1088/1742-6596/1154/1/012002

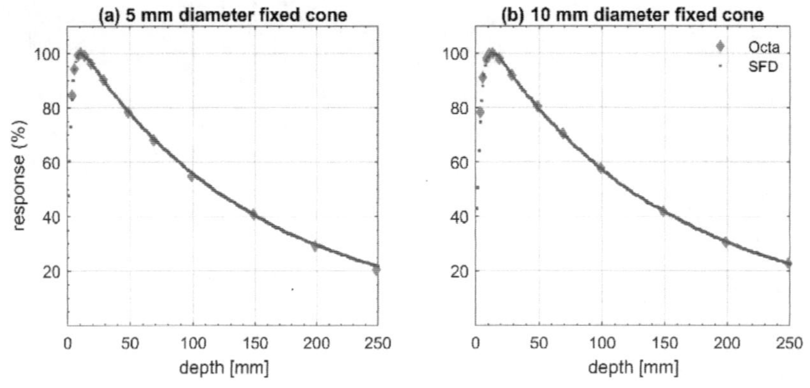

Figure 4. PDD measured by the Octa and a SFD diode for a (a) 5 mm and (b) 10 mm diameter field collimated by a fixed conical cone.

4. Discussion and Conclusion

An accurate measurement of OFs for small radiation fields is paramount. Their high sensitivity to small changes in cone size, due for instance to manufacture-related variations, was demonstrated, with a difference in OF of about 10% for 0.3 mm variation in a 5 mm cone aperture [12]. Silicon diodes are known to require corrections for OF measurements due to electron spectra perturbations with respect to water [13,14]. In this study, OFs measured by the Octa were accurate within 3% with respect to those measured by EBT3 films, a result which supports the current 'correction-free' design of the detector for these measurement conditions.

For fixed cones, DPs by the Octa were overall in good agreement with the EBT3 measurements in terms of FWHM, with discrepancies well within 3% for fields equal to or larger than 10 mm in diameter, and approximately 5% for the 5 mm and 7.5 mm field diameters. A maximum discrepancy of 6.9% or 0.4 mm was found for the in-line profile of the 5 mm cone. The elliptical shape of the electron source in Elekta linacs partially explains the expected small differences between in- and cross-plane penumbra widths [12]. The Octa measured a penumbra width of 1.8 mm for the in-plane profile and of 2.1 mm for the cross-plane profile. Overall, penumbra widths measured by the Octa were generally in close agreement with those measured by EBT3 films, with discrepancies within 0.3 mm.

For PDD measurements, for silicon detectors a decrease in sensitivity is expected with decreased dose per pulse [15]. This could be partially offset by an overestimation of the dose due to the increase of the relative number of low energy scattered photons with increasing depth [16,17]. In this study, while no correction was made to account for dose rate response variations, discrepancies in PDD measured by the Octa with respect to the benchmarks were within 3% at all depths, for all field sizes investigated. The gantry sag was determined to be minimal (within 0.3 mm) in the cross-plane direction and approximately 1.2 mm in the in-plane direction.

Overall, the Octa was demonstrated to be an accurate 'correction-free' detector for small field dosimetry with potential for the dosimetric characterization of stereotactic dedicated collimators such as fixed cones. Allowing for the simultaneous real-time read-out of OF and DPs along cross- and in-plane along with 2 diagonal directions, all acquired with sub-mm accuracy for any given field size, the Octa would greatly reduce the measurement time needed to comply with quality assurance protocols.

5. Acknowledgments

We would like to acknowledge the Gross Foundation for financial support and the Prince of Wales Hospital for access to their linac.

6. References

[1] ICRU 2017 *ICRU Report 91*
[2] IAEA 2017 *IAEA-TRS-483*
[3] Gorjiara T *et al* 2011 *Phys. Med. Biol.* **56** 4685-99
[4] Vial P *et al* 2008 *Med. Phys.* **35** 4362-74

[5] Benmakhlouf H and Andreo P 2017 *Med. Phys.* **44** 713-24
[6] Francescon P *et al* 2017 *Phys. Med. Biol.* **62** 1076-95
[7] Charles P H *et al* 2014 *Med. Phys.* **41** 101701
[8] Huet C *et al* 2016 *Phys. Medica* **32** 818-25
[9] Biasi G *et al* 2018 *Phys. Medica* **45** 117-26
[10] Biasi G *et al* 2018 *J. Appl. Clin. Med. Phys.* **19** 547-57
[11] Aldosari A H *et al* 2014 *Med. Phys.* **41** 091707
[12] Borzov E *et al* 2018 *J. Appl. Clin. Med. Phys.* **19**194-203
[13] Andreo P 2018 *Radiother. Oncol.* **126** 205-13
[14] Looe H K *et al* 2018 *Phys Med Biol* **63** ***015001***
[15] Rikner G and Grusell E 1983 *Phys. Med. Biol.* **28** 1261-7
[16] Zhu X R *et al* 2000 *Med. Phys.* **27** 472-7
[17] Eklund K and Ahnesjö A 2009 *Phys. Med. Biol.* **54** 7411-23

IOP Conf. Series: Journal of Physics: Conf. Series **1154** (2019) 012003 doi:10.1088/1742-6596/1154/1/012003

Modelling the Biological Beamline at HIMAC using Geant4

D Bolst[1], L T Tran[1], S Guatelli[1], N Matsufuji[2] and A B Rosenfeld[1]
[1]Centre for Medical Radiation Physics, University of Wollongong, Australia
[2]Research Centre for Charge Particle Therapy, National Institute of Radiological Science, Chiba, Japan

E-mail: db001@uowmail.edu.au

Abstract. A Geant4 simulation modelling the passive Biological beamline at the Heavy Ion Medical Accelerator Chiba (HIMAC) is described and validated against experimental measurements for a mono-energetic and spread out Bragg peak (SOBP) ^{12}C beam. Comparisons to experimental data showed good agreement for both lateral and depth dose profiles. It was found that simplifying the simulation model by generating a cone beam instead of using the wobbler magnets provided a good approximation for both lateral and depth dose measurements.

1. Introduction

The Heavy Ion Medical Accelerator (HIMAC) at the National Institute of Radiological Sciences (NIRS), in Chiba, Japan, was the first centre to clinically treat patients with ^{12}C and as of 2015 has treated more than 10000 patients with ^{12}C therapy [1]. Since then, ^{12}C therapy has had a growing interest thanks to its enhanced physical and biological dose properties, however, the ^{12}C beam produces a complex radiation field which needs to be fully characterised for Quality Assurance. Monte Carlo simulations are a very useful tool supporting the research in ^{12}C ion therapy, however, simulations must be validated against experimental data to quantify their predictive accuracy. In this work the modelling of the Biological beamline at HIMAC using Geant4 is described and validated against experimental measurements. This simulation model is used at the Centre for Medical Radiation Physics, University of Wollongong, to study novel detector solutions for Quality Assurance including silicon micro dosimeters [2-4].

2. Materials and Methods

2.1. Simulation Setup

The Biological beamline was modelled in Geant4 [5, 6] using version 10.2p2. The layout of the modelled beamline can be seen in figure 1. The ^{12}C beam is generated at the beam duct in a vacuum tube 205 cm in length. The ^{12}C beam has an initial energy of 290 MeV/u and a 0.2% energy σ. The beam is generated as a Gaussian distribution with a spot size of 3.4 mm σ.

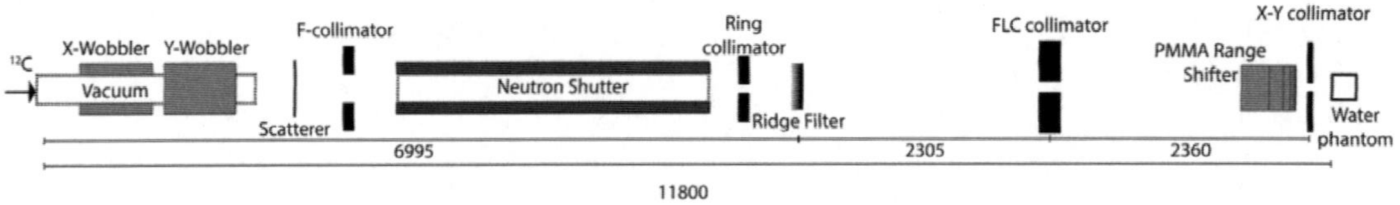

Figure 1. Layout of the Biological beamline modelled in Geant4, not to scale.

The beam is then shaped to a circle by two wobbler magnets operated using the single-ring wobbling method [7]. The strength of the magnetic field at a particular time t is $B_0 \sin(\omega t)$ and $B_0 \cos(\omega t)$ for the X and Y wobbler magnets respectively, where ω is 56.41 Hz. In the simulation the period is sampled at 111 different times. Each wobbler magnet is modelled as a cylinder with a radius of 250 mm and length of 660.1 mm with a 110 mm gap between the two wobbler magnets. In the simulation a uniform magnetic field is applied to both the wobbler magnets.

Once the beam is shaped by the pair of wobbler magnets the lateral dose uniformity of the beam is improved by passing through a tantalum scatterer which is located at ~2308 mm from the beam duct. The thickness of the scatterer and wobbler strength, B_0, depend on the energy and ion of the beam being delivered, as well as the field size. For the mono-energetic 290 MeV/u ^{12}C beam the B_0 is 0.045 T and the scatterer is 0.434 mm thick while the SOBP has a B_0 of 0.061 T and a scatterer thickness of 0.649 mm.

After passing through the scatterer, the beam is collimated by a 100 mm thick brass F-collimator with a diameter of 100 mm, located 2755 mm from the beam duct. After the F-collimator the contribution of neutrons in the beam is reduced by traversing a 2885 mm long vacuum neutron shutter. After exiting the neutron shutter, the beam is further collimated by a 200 mm thick brass ring collimator with a diameter of 100 mm.

In the SOBP configuration, the beam will then pass through a ridge filter (RGF), located 6995 mm from the beam duct. For a 6 cm 290 MeV/u ^{12}C SOBP the RGF is made up of 21 bars of milled aluminium 200 mm in length and ~5 mm wide, each bar consists of 101 different thicknesses.

The beam then traverses a 20 cm thick aluminium, four leaf collimator (FLC) placed 9300 mm from the beam duct. The beam can then pass through various combinations of PMMA thicknesses to shift the range of the beam, though none are used in this study. The beam can then be further collimated by a 5 cm thick X-Y brass collimator, 140 mm before isocentre.

The changing shape of the beam as it traverses the beamline is shown in figure 2, which shows the transport of ~10^4 mono-energetic ^{12}C ions from the beam duct to the isocentre with a field size of 100×100 mm^2 set by the collimators.

In the Geant4 simulation after the final brass collimator there is the option to generate a phase space file (PSF) by recording particle details in a ROOT file such as the particle type (ion, neutron, electron or photon). If the particle type is an ion, then the A and Z will also be stored. Each particle's x, y, z position and momentum are also stored along with its kinetic energy. When reading in the PSF, 1000 entries are read into memory at a time to increase the speed of the simulation.

In this study two phantom sizes were used for scoring the lateral and depth doses with sizes of 30×30×30 cm^3 and 22×22×30 cm^3 respectively, where 30 cm is the thickness for both cases. The phantom size of 22×22 cm^2 was used to match the size used in [8]. The scoring of the lateral dose profile was done in 5×5×10 mm^3 voxels at the surface of the phantom. For calculating the depth dose, the phantom had 1 mm thick water cylinders with diameters of 10 mm placed along the beam axis.

There are various 10 μm thick aluminium sheets placed throughout the beamline. These represent ionisation chamber (IC) monitors and the end of the first vacuum chamber and the start and end of the neutron shutter. The IC are made up of two sheets of aluminium and are positioned at 2175, 6910, 6950 and 8525 mm from the beam duct.

To describe electromagnetic interactions the *G4StandardOption3* physics list was adopted. Hadronic interactions were modelled using the Binary Intranuclear Cascade (BIC) which has been recently benchmarked against other alternative models in Geant4 [9] for a therapeutic ^{12}C beam. The elastic scattering of hadrons was modelled by means of the *G4HadronElasticPhysicsHP* and the neutron High Precision (HP) model was adopted to describe neutron interactions of energies up to 20 MeV. Within the water phantom a production cut size and step limit of 0.5 mm were used.

In order to validate the Geant4 simulation both lateral and depth dose profiles were compared to experimental measurements. For a 290 MeV/u mono-energetic and 6 cm SOBP ^{12}C beam experimental data from [10] and [11] were used, respectively, for the lateral dose. For both these cases the dose is measured at the surface of the phantom with an ionisation chamber with the FLC and brass collimator being fully opened. For depth dose measurements data from [8] were used for both beams. For lateral dose comparisons the maximum value in the experiment and simulation were both normalised to the

IOP Conf. Series: Journal of Physics: Conf. Series **1154** (2019) 012003 doi:10.1088/1742-6596/1154/1/012003

same value, while for reference depth dose comparisons the first depth in the experiment was used to normalise the experiment and simulation distributions.

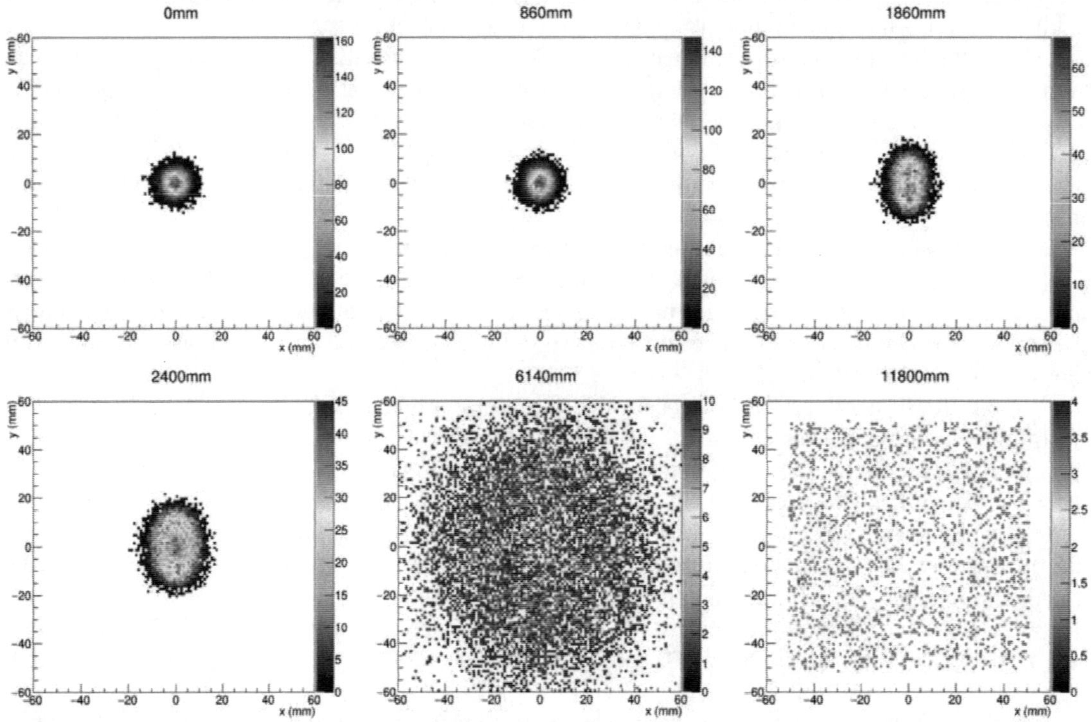

Figure 2 A beam's eye view of the fluence of a mono-energetic ^{12}C beam with ~10^4 primary particles at various distances from the beam duct. The 860 and 1860 mm positions correspond to the exit of the first and second wobbler magnets, respectively. The 2400 and 6140 mm distances show the profile of the beam after traversing the scatterer and neutron shutter, respectively. The 11800 distance is the profile at isocentre after being collimated to a 100×100 mm^2 size.

To investigate whether the simulation could be simplified by excluding the wobbler, the lateral and depth dose profiles were compared to the case of having the wobbler setup with the FLC and brass collimators set to a field size of 100 × 100 mm^2, a common field size of interest. The simplified methods consist of deactivating the wobbler magnets and generating the particles at the same beam duct position as used for the full wobbler technique. The different methods included simulating a 3.4 mm σ pencil beam (the same initial beam size used for the full wobbler method), a cone beam simulated by generating an isotropic distribution with an aperture angle of 1 degree and finally a circular beam with diameter of 100 mm and generated uniformly. A beam's eye view of the fluence using the different methods is shown in figure 3, the positions are at the beam duct where the beam is generated (0 mm) and just before the neutron shutter (3200 mm).

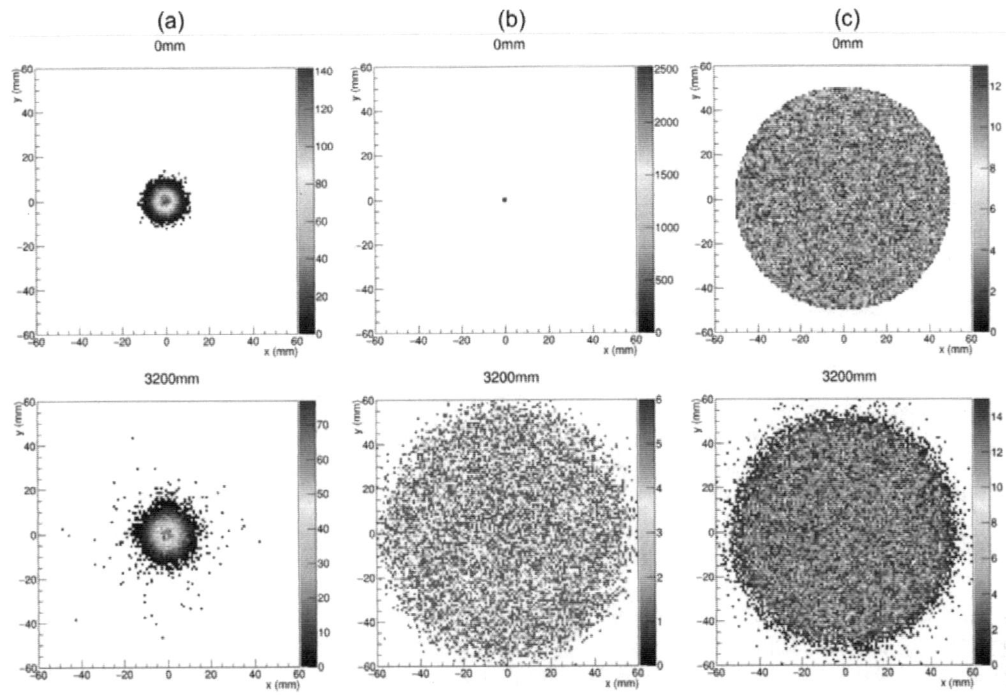

Figure 3. A beam's eye view of the fluence of the beam simulated with different methods at the beam duct (0 mm) and just before the neutron shutter (3200 mm). (a) 3.4 mm σ pencil beam, (b) 1° cone beam and (c) 100 mm diameter circle beam.

3. Results and Discussion

3.1. Validation

Figure 4 shows the comparison of simulation and experimental data in terms of the lateral and depth dose profiles with the percentage difference (PD) plotted at the bottom. The PD was calculated using equation (1). Overall, the agreement between simulation and experiment is satisfactory for both the lateral and depth dose profiles.

$$PD = \left(\frac{Exp - Sim}{Exp}\right) \times 100\% \quad (1)$$

The lateral dose profiles at the surface of the phantom are shown for a mono-energetic beam (a) and a SOBP (b). It can be seen that for the pristine beam there is excellent agreement between the experiment and simulation with a difference of less than 0.5% in the treatment field between ±50 mm. At the sharp penumbra the difference increases to a maximum of ~2.5%, however due to the rapidly changing dose in this region experimental positioning becomes more sensitive. For the SOBP there is visibly less agreement, with the treatment field varying between ~2%, however the penumbra gives agreements within ~1%.

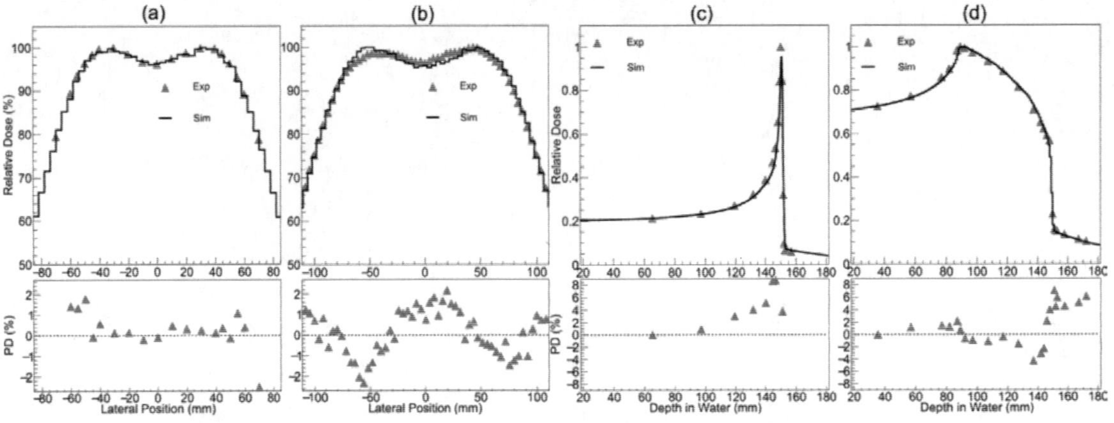

Figure 4. Comparison between the relative dose calculated by means of the simulation and experiment with the percentage difference (PD) plotted on the bottom. Plots (a) and (b) show the lateral dose at the surface of the phantom for a mono-energetic (a) and 6 cm SOBP (b) 290MeV/u ^{12}C beam. Plots (c) and (d) compare the depth dose distribution for the mono-energetic (c) and SOBP (d) beams.

The depth dose distributions for a mono-energetic and SOBP 290 MeV/u ^{12}C beam are shown in (c) and (d), respectively. Both setups are done using a collimation size of 100×100 mm^2 for both the FLC and brass collimator. For the mono-energetic beam the overall shape agrees very well, however the difference between the simulation and the experiment gradually increases with depth and has a maximum of ~8% at the pinnacle of the BP. The SOBP depth dose (d) shows a more stable difference between the simulation and the experiment with a difference between ~2%. Like with the mono-energetic beam the SOBP beam also has a maximum difference at the end of the SOBP range, these regions have sharp dose gradients and are sensitive to slight misplacements or energy differences of the delivered beam.

3.2. Simplified simulations techniques to model the effect of the wobbler

A comparison of the different simplified simulation techniques is shown in figure 5, with the plotted percentage differences being calculated with respect to the full wobbler method.

The lateral dose is compared using a mono-energetic beam collimated to a 100×100 mm^2 field and is shown in (a). It can be seen that the cone beam reproduces the lateral beam profile of the wobbler setup very well, with a difference within ~2% for the treatment field with the difference increasing at the penumbra, while both the point and circle beams agree very poorly.

The depth dose, using the different simulation techniques is shown in (b) for the mono-energetic beam and (c) for the SOBP. For the mono-energetic beam the different methods of modelling the beam do not vary strongly with one another, with the cone beam not differing significantly. The pencil and circle beams are very similar to one another, with them both increasing in difference with depth compared to the wobbler method, with a difference of ~1% higher at the pinnacle of the BP, after the BP they both under respond compared to the wobbler with a difference which increases with depth.

For the case of the SOBP (c), shows a much more prominent difference between the cone beam compared to the point and circle beams. Without the angular spread, the point and circular beams create a flatter dose over the range of the SOBP between 85 and 145 mm, with a maximum difference of ~15% at the end of the SOBP.

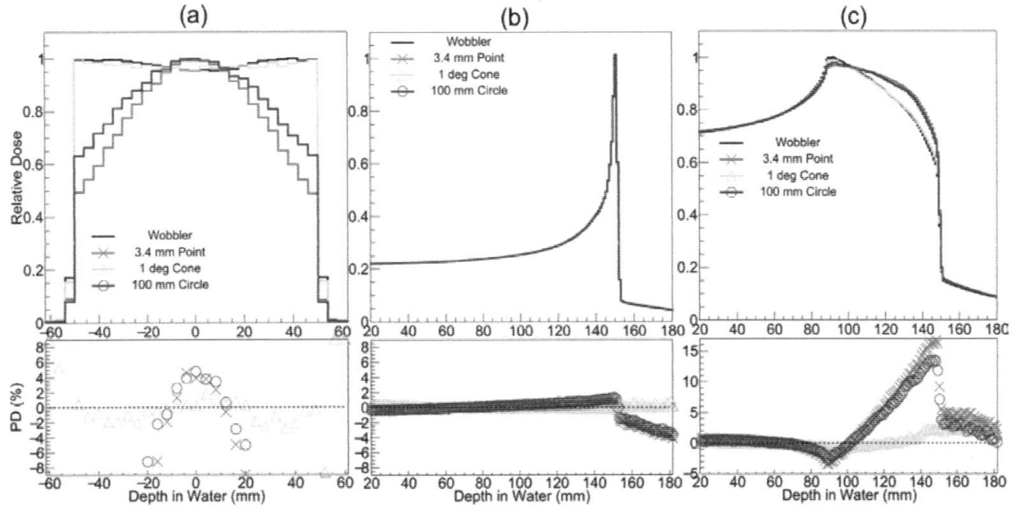

Figure 5. Comparison of the lateral dose for a mono-energetic beam (a) and the depth dose of a mono-energetic (b) and a SOBP (c) when using different methods to model the effect of the wobbler on the beam shape. The percentage difference (PD) is calculated with respect to the dose distributions using the wobbler simulation.

4. Conclusion

A Geant4 application modelling the Biological Beamline at HIMAC, NIRS, has been described and validated against experimental measurements for a mono-energetic and SOBP 290 MeV/u ^{12}C beam. Comparisons with experiment for the lateral beam profile showed good agreement for both mono-energetic and SOBP beams, with differences of ~0.5% and 2% in the treatment field, respectively. For the depth dose profiles, both beams gave good agreement in the overall shape, however both have a maximum difference with the experiment at the end of the BP where measurements are especially sensitive. In order to simplify the simulation by excluding the wobbler system a number of alternative methods were modelled. It was found that for both lateral and depth dose profiles simulating a cone beam provided very good agreement compared to modelling the full wobbler system.

5. References

[1] https://www.ptcog.ch/
[2] Bolst D *et al* 2017 *Rad. Meas. A* **106** 512-518
[3] Chartier L *et al* 2017 *Rad. Prot. Dos.* **180** 365-71
[4] Tran L *et al* 2018 *Med. Phys.* **44** 6085-6095
[5] Agostinelli A et al 2003 *Nucl. Instr. Meth. Phys. Res. A* **506** 250-303
[6] Allison J et al 2016 *Nucl. Instr. Meth. Phys. Res. A* **835** 186-225
[7] Yonai S *et al* 2008 *Med. Phys.* **35** 2473-4209
[8] Tran L *et al* 2018 *Radiat. Meas.* **115** 55-59
[9] Bolst D et al 2017 *Nucl. Instr. Meth. Phys. Res. A* **869** 68-75
[10] http://wrmiss.org/workshops/sixteenth/Ploc.pdf
[11] Yonai S *et al* 2009 *Med. Phys.* **36** 927-938

IOP Conf. Series: Journal of Physics: Conf. Series **1154** (2019) 012004 doi:10.1088/1742-6596/1154/1/012004

Microdosimetry on nanometric scale with a new low-pressure avalanche-confinement TEPC

D Bortot[1,2,4], D Mazzucconi[1,2], S Agosteo[1,2], A Pola[1,2], S Pasquato[1,2], A Fazzi[1,2], P Colautti[3] and V Conte[3]

[1]Politecnico di Milano, Energy Department, via La Masa 34, 20156 Milano, Italy

[2]INFN, Sezione di Milano, via Celoria 16, 20133 Milano, Italy

[3]INFN, Laboratori di Legnaro, viale dell'Università 2, Legnaro, Padova, Italy

E-mail: davide.bortot@polimi.it

Abstract. The tissue equivalent proportional counter (TEPC) is the most accurate device for measuring the microdosimetric properties of a particle beam, nevertheless no detailed information on the track structure of the impinging particles can be obtained, since the lower operation limit of common TEPCs is about 0.3 μm. On the other hand, the pattern of particle interactions is measured by track-nanodosimetry, which derives the single-event distribution of ionization cluster size at the nanometric scale. Anyway, only three nanodosimeters are available worldwide. A feasibility study for extending the performances of TEPC down to the nanometric region was performed and a novel avalanche-confinement TEPC was designed and constructed. This detector is constituted by a cylindrical chamber, based on a three-electrode structure, connected to a vacuum and gas flow system to ensure a continuous replacement of the tissue equivalent gas, thus allowing to simulate different biological site sizes in the range 300-25 nm. This TEPC can be calibrated by exploiting a built-in alpha source and a miniaturized solid-state detector as a trigger. Irradiations with photons, fast neutrons and two hadron beams demonstrated the good performances of the device. A satisfactory agreement with FLUKA simulations was obtained.

1. Introduction

The interest in hadron therapy, based mainly on protons and carbon ions for treating radio-resistant cancers and malignant tumours close to critical organs, has been growing during the last years [1]. Ion beams provide a better dose conformation and show superior radiobiological properties with respect to conventional radiation therapy. Since ionization generated in ion therapy is markedly localized and non-uniform against depth, significant variations on the radiation quality and consequently differences in the biological effectiveness across the Bragg curve are present [2]. Nevertheless, the present radiation- treatment planning procedures are based on the measurement of the absorbed dose, which is a macroscopic and average quantity that does not take into account neither the stochastics of particle interactions in the target volume nor the track structure of ionizing charged particles, which is crucial for the initiation of the radiation damage [3]. The standard dosimetric approach can be integrated with methodologies and instruments provided by microdosimetry, which aims at measuring the statistical fluctuations of the local energy imparted at the micrometric level, and track-nanodosimetry, which is devoted to the description of the pattern of particle interactions in nanometric volumes.

The tissue equivalent proportional counter (TEPC) is the most accurate device for measuring the microdosimetric properties of a particle beam, but it is capable of simulating site sizes in the

IOP Conf. Series: Journal of Physics: Conf. Series **1154** (2019) 012004 doi:10.1088/1742-6596/1154/1/012004

micrometric domain down to about 0.3 μm when operated in pulse-height mode [4]. Nevertheless, it is widely accepted today that radiation damage is initiated at the DNA level and the severity of the damage is closely related to the radiation track structure, which causes inhomogeneous interaction patterns at the nanometric scale [5]. Probability distributions of such interaction patterns, which are measured by track-nanodosimetry, show a trend similar to the cellular inactivation cross-sections [6]. However, only three nanodosimeters are available worldwide, showing stringent limitations: complexity, dimension and associated lack of transportability. This technological hurdle could be partially overcome by developing a microdosimeter capable of simulating tissue sites down to the nanometric region. Since the lower operation limit of single-wire TEPCs is about 300 nm in order to maintain an acceptable energy resolution, it is necessary to modify the geometry of the sensitive volume by embedding a third electrode for confining the electronic avalanche within a defined region. An extensive study, based on a prototype described elsewhere [7], allowed designing and developing a new low-pressure avalanche-confinement TEPC capable of measuring microdosimetric distributions in simulated sites in the range 300-25 nm.

2. The avalanche-confinement TEPC

The cylindrical sensitive volume of the new avalanche-confinement TEPC (13 mm in diameter and length) houses three electrodes biased independently: a central anode wire (graphite), a cylindrical cathode shell (conductive plastic A-150 type) and a helix (gold-plated tungsten), which surrounds the anode and subdivides the sensitive volume into an external drift zone and an internal multiplication region. Two aligned cavities embed, respectively, a removable ^{244}Cm alpha source and a very compact solid-state detector: this configuration allows calibrating the TEPC by also varying the simulated site size and the polarization of the three electrodes (Figure 1). It guarantees that only signals due to alpha particles with a straight path inside the sensitive volume, i.e. the drift region, are collected [8]. A customized and transportable vacuum and gas flow system guarantees vacuum conditions and ensures a continuous replacement of tissue equivalent gas inside the chamber. Dimethyl ether (DME: $(CH_3)_2O$), which can be considered as a tissue-equivalent gas apart from the lack of nitrogen, is the selected filling gas for this TEPC.

Irradiations with photons emitted by a ^{137}Cs source confirmed that the operating parameters of the device, i.e. the bias voltages of the electrodes, do not affect the microdosimetric information measured by the detector, which depends on the cavity size and on the radiation field only. Moreover, the TEPC response in the range 0.3 μm-25 nm against a fast neutron field produced by a calibrated ^{241}Am-Be source and quasi-monoenergetic neutron beams produced through the ^{7}Li(p,n)^{7}Be reaction on a LiF target was assessed experimentally. The comparison between measured distributions and FLUKA simulated spectra shows a rather good agreement also at low site sizes [9].

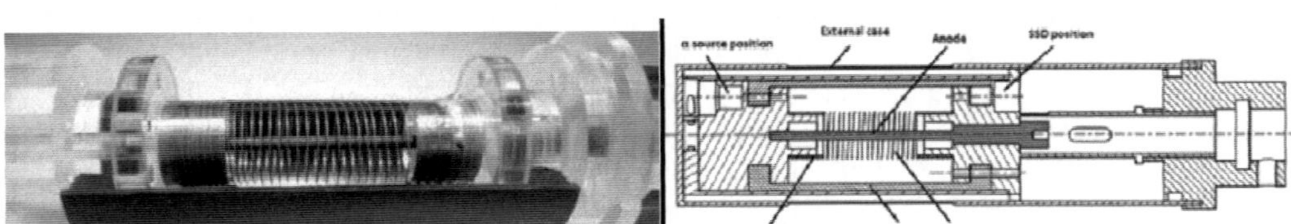

Figure 1. The avalanche-confinement TEPC: the calibration alpha source and the solid-state detector (SSD) are indicated in the cross-sectional view.

3. Response against 62 MeV/u carbon ion and helium ion beams

The avalanche-confinement TEPC was characterized preliminarily against a 62 MeV/u carbon ion beam. A set of PMMA foils with different thicknesses was used to reproduce different depths across the Bragg peak. Several configurations were tested by varying both the simulated site size in the range 300-25 nm and the depth across the Bragg peak of the delivered beam [10]. Figure 2 shows the dose distributions obtained for a simulated site of 300 nm at a depth 6.50 mm, 6.76 mm and 7.09 mm in

PMMA. At 6.50 mm the carbon peak is slightly shifted at lower lineal energy values with respect to the one at 6.76 mm, but the edge is the same. At 7.03 mm, the spectrum changes completely: the primary carbon contribution is no more present and that due to fragments, which is barely visible at 6.50 mm and 6.76 mm, becomes predominant. This depth is beyond the Bragg peak. The microdosimetric spectra obtained at eight different points across the Bragg peak are shown in Figure 3 for a 25 nm simulated site. The distributions shift towards higher *y* values as depth increases (from 1.76 mm to 6.76 mm), according to the beam slowing-down. In contrast, at 7.29 mm in depth, which is downstream of the Bragg peak, the distribution changes dramatically due to the contribution of fragments. The experimental set-up was also simulated with Monte Carlo simulations performed with the

FLUKA code in order to compare experimental and simulated microdosimetric spectra in site sizes equal to 300 and 25 nm. The obtained results are in good agreement, showing that this code is capable of reproducing microdosimetric spectra of a carbon beam down to 25 nm in simulated site [11]. Nevertheless, it should be stressed that these simulations refer to carbon beam only and further comparison with experimental data measured with other particles is required.

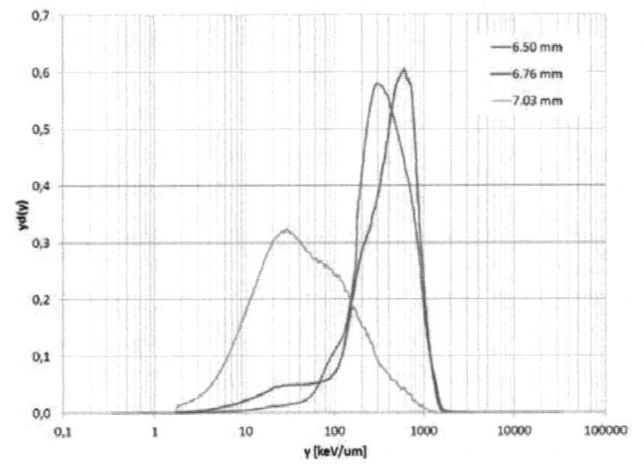

Figure 2. Carbon microdosimetric spectra obtained at 6.50, 6.76 and 7.03 mm in PMMA for a 300 nm simulated site.

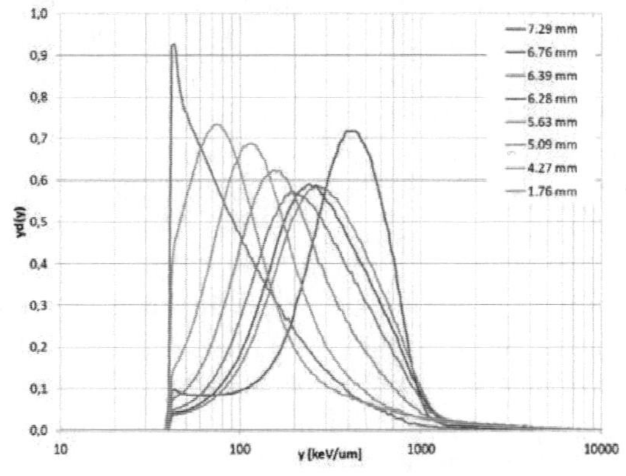

Figure 3. Carbon microdosimetric spectra obtained at different depths in PMMA from 1.76 mm to 7.29 mm for a 25 nm simulated site.

A very recent characterization was performed by irradiating the TEPC with a 62 MeV/u alpha particle beam. The experimental response of the microdosimeter for different simulated site sizes in the range 300-25 nm at seven points across the depth dose distribution was assessed. Figure 4 shows the microdosimetric spectra obtained for a 300 nm site at five different depths in PMMA. The distributions shift towards higher lineal energy values as depth increases, according to the beam slowing-down. In Figure 5, the comparison of the spectra obtained at different simulated site sizes at the same depth (24.31 mm in PMMA) is shown. The alpha edge does not change position, while a slight shift of the main peak towards higher lineal energy values can be observed with decreasing the site size. FLUKA simulations of all studied irradiation conditions are still ongoing, but a first good agreement was obtained for the 300 nm site size.

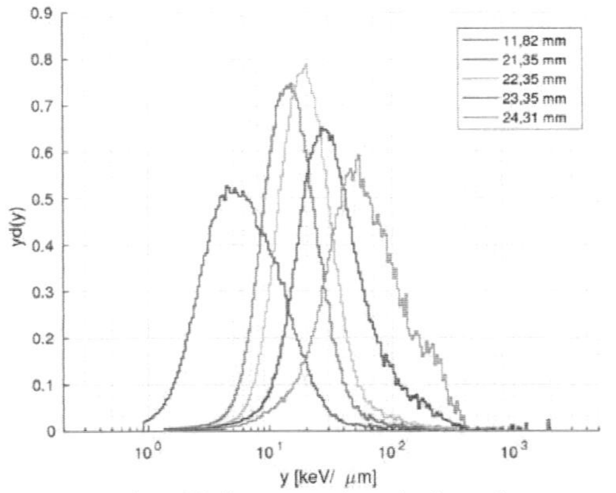

Figure 4. Helium microdosimetric spectra obtained at different depths in PMMA for a 300 nm simulated site.

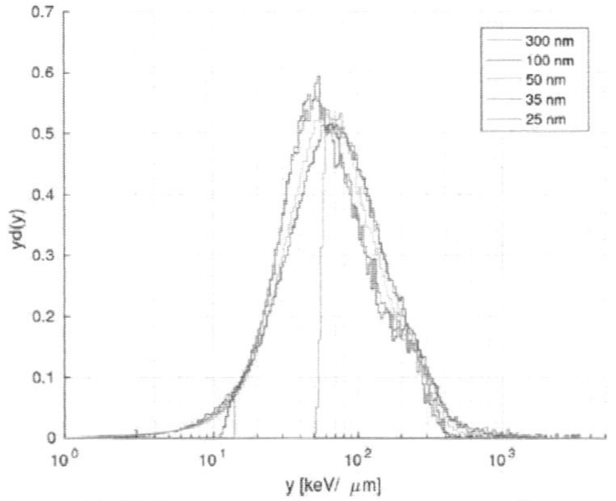

Figure 5. Helium microdosimetric spectra for different simulated site sizes in the range 300-25 nm at 24.31 mm in PMMA.

4. Conclusions

The irradiation campaigns with different neutron beams and low-energy hadrons (helium and carbon ions) give confidence about the capability of this novel avalanche-confinement TEPC of measuring microdosimetric distribution at simulated site ranging from 0.3 μm down to 25 nm. Nevertheless, further irradiations with other particles are necessary to study deeply the charge collection efficiency of the TEPC at low simulated site sizes (down to 25 nm). Moreover, further comparisons between FLUKA simulation and experimental data measured with other particles are foreseen.

5. Acknowledgments

This work was supported by the Italian National Institute for Nuclear Physics - INFN – Scientific Commission V in the framework of the MITRA (Microdosimetry and TRAck structure) and NADIR (biologically relevant NAnoDosimetry of Ionizing Radiation) projects.

6. References

[1] Durante M and Loeffler J S 2010 *Nat. Rev. Clin. Oncol.* **7** 37-43
[2] Gorjiara T *et al* 2012 *Med. Phys.* **39** 7071-9
[3] Weyrather K and Kraft G 2004 *Radioth. Oncol.* **73** S161-9
[4] Hogeweg B 1973 *Proc. 4th Symp. on Microdosimetry* **5122** pp.843-854
[5] Lomax M E *et al* 2013 *Clin. Oncol.* **25** 578-85
[6] Conte V *et al* 2017 *Radiat. Prot. Dosim.* **180** 150-6
[7] Cesari V *et al* 2002 *Radiat. Prot. Dosim.* **99** 337-42
[8] Bortot D *et al* 2017 *Radiat. Meas.* **106** 531-7
[9] Bortot D *et al* 2017 *Radiat. Prot. Dosim.* **180** 172-6
[10] Bortot D *et al* 2017 *Radiat. Meas.* **103** 1-12
[11] Mazzucconi D *et al* 2018 *Radiat. Meas.* **113** 7-13

IOP Conf. Series: Journal of Physics: Conf. Series **1154** (2019) 012005 doi:10.1088/1742-6596/1154/1/012005

Brachy*View*: verification of LDR patient plans – hardware optimisation

T Brennen[1], D L Cutajar[1,2], S Alnaghy[1], J Bucci[2], A Bece[2], M Favoino[3], M Carriero[3], M Lerch[1], A B Rosenfeld[1] and M Petasseca[1]
[1]Centre for Medical Radiation Physics, University of Wollongong, Australia
[2]St George Cancer Care Centre, St George Hospital, Kogarah, Australia
[3]Advanced Computer System Pty Ltd, Rome, Italy

E-mail: tjb790@uowmail.edu.au

Abstract. BrachyView is a novel in-body imaging system developed with the objective to provide real-time intraoperative dosimetry for low dose rate (LDR) prostate brachytherapy treatments. Seed positions can be reconstructed after implantation by the means of a high-resolution pinhole gamma camera. The obtained dataset is then combined with conventional trans-rectal ultrasound (TRUS) imaging to locate the effective source position within the prostate volume. This study presents a comparison of two data sets containing 96 and 98 I-25 brachytherapy sources and readout by two different pin-hole collimators with 800 μm and 500 μm, respectively. This study aims to assess the effect on reconstruction of large pinholes which allow for large Field of View (FoV) and statistics of counts on the seed's projections. It was found that 100% of seeds were reconstructed by the means of the 500 μm collimator. Results from the 800 μm data showed that only 70% of seeds were reconstructed with a total 3D discrepancy of 8.20 mm in respect to CT data and an increased reconstructed position inaccuracy.

1. Introduction

Prostate cancer is a disease by which malignant cells form in the tissues of the prostate gland. Prostate cancer is quickly becoming the most common form of cancer across the globe. Prostate cancer is one of the most commonly diagnosed cancers within Australia, contributing to 24.7% of newly diagnosed cancers [1]. With the current aging population this problem is predicted to continue to rise within the following years, as prostate cancer is most common in elderly males, with 85% of prostate cancer cases being attributed to men above the age of 65 [2]. It is hence of grave importance that common treatment modalities available for prostate cancer treatment can be developed and improved in order to support patients and provide the best care possible.

Common treatments prescribed to patients include radical prostatectomy (RP), involving the removal of the entire prostate gland, external beam radiation therapy (EBRT) consisting in delivery of focused x-ray beams to the treatment area, and finally brachytherapy treatments, either consisting of low dose rate (LDR) or high dose rate (HDR) treatments [3]. Radiation oncologists must carefully evaluate the patient in order to prescribe the most suitable treatment, some clinical measures for determination of patient selection for a treatment includes the grade of the tumour, Gleason score and prostate specific antigen scores (PSA) [4].

LDR brachytherapy is a radiotherapy treatment that is most commonly utilized to treat low rade, slowly proliferating tumours located within the prostate. Permanent radioactive implants are used to deliver the prescribed dose to the cancerous area, these implants are commonly referred to as seeds and contain I-125 or Pd-103. The use of these specific radionuclides is to ensure a highly conformal dose is delivered to the tumour site, with steep dose gradients utilized to minimize the dose delivered to surrounding healthy tissues and critical organs such as the bladder, urethra and rectum [5]. It is critically important to minimize the does to surrounding healthy tissues as additional dose can lead to poor patient outcomes with post-operative complications such as sexual dysfunction and urinary symptoms, impacting the quality of life of the patient.

Studies have shown that additional dose to surrounding healthy tissues can results from seed positioning errors during the implantation procedure [6]. Due to possible seed position deviations from those outlined in the treatment planning system it is crucial to re-assess the true positions of the seeds directly after the implantation. Such procedures require the determination of the seeds positions and the resulting dose distribution to be correctly determined in real time and in-vivo. Currently, patient doses are verified utilizing means of post-operative dosimetry such as computerised tomography (CT) or magnetic resonance imaging (MRI) to visualize and reconstruct seed positions [7]. However, this procedure is commonly performed a number of weeks or even months after the procedure. As the desire to optimize and improve the overall treatment of brachytherapy becomes ever more important it is crucial to ensure that a means for real-time intraoperative dosimetry be developed, allowing for real-time localization of seeds with no additional dose delivered to the patient, allowing for intraoperative replanning to ensure adequate coverage of the planned treatment volume and minimization of dose to the organs at risk. CMRP propose the system named BrachyView, the first miniaturised gamma camera able to determine the seeds positions in-vivo and in real-time.

2. Materials and Methods

The BrachyView probe consists of a 1 mm thick cylindrical tungsten collimator, containing three single cone pinholes, as shown in Figure 1. This study utilized two collimators with pinhole diameters of 500 μm and 800 μm to assess the effect of the pinhole diameter on seed reconstruction quality in the presence of a clinically relevant number of seeds, as well as the feasibility to perform reconstruction prior to seed deployment from the treatment needles. Details on the probe architecture and construction structure has been previously presented by Alnaghy [8].

Figure 1. The BrachyView probe covered in a 700 μm kapton tape, in order to ensure no liquid penetration during acquisition.

Two clinical LDR prostate brachytherapy plans were devised containing 98, and 96 seeeds (I-125 with an average activity of 0.248mCi and 0.303mCi respectively), implanted into a CIRS prostate gel phantom. TRUS images, manual segmentation and rendering were performed to reconstruct the 3D shape and position of the prostate, utilizing transversal 2.5 mm ultrasound slices.
BrachyView data was acquired for an average of 2.5 minutes after each seed deployment to compensate for the low activity of the seeds. A post implant CT study with O-MAR (orthopedic metal artefact reduction) and scan protocol of 1 mm slice thickness with a 0.5 mm inter-slice acquisition, was completed to provide a reference for comparison of the reconstructed seed positions.

IOP Conf. Series: Journal of Physics: Conf. Series **1154** (2019) 012005 doi:10.1088/1742-6596/1154/1/012005

3. Results

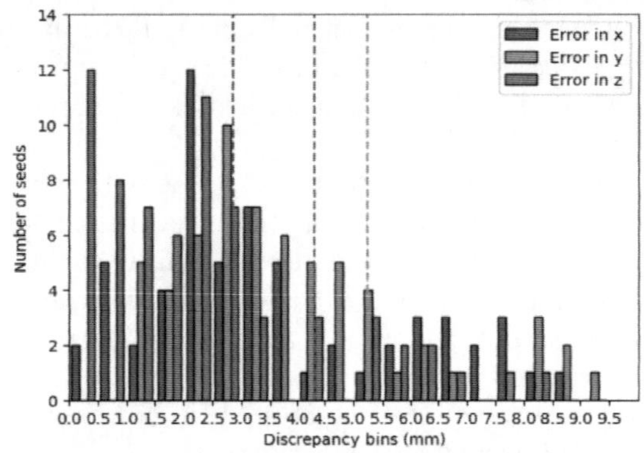

(a)

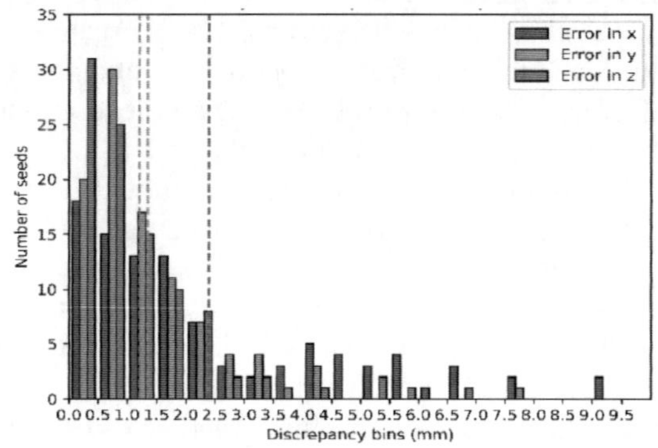

(b)

Figure 2. a) CT and BrachyView 800 μm collimator, **b)** 500 μm collimator reconstructed seed position discrepancies with average discrepancies denoted as dotted lines.

When compared to the seed positions obtained by the CT scan the BrachyView reconstruction showed average discrepancies of 2.4 mm, 1.35 mm and 1.2 mm in the x, y and z reference frames, respectively for the 500 μm collimator. For the 800 μm collimator average discrepancies of 4.3 mm, 5.23 mm and 2.86 mm in the x, y and z reference frames respectively. Only 70% of seeds were reconstructed, when compared to 100% of seeds reconstructed with use of the 500 μm collimator.
The use of the 800 μm collimator leads to larger X, Y and Z discrepancies in respect to the reference data from CT than the 500 μm collimator, as shown in Figure 2. This can be attributed due to the increase in scatter contribution within the images, as well as an increase in projection size during image acquisition. Projection size plays an important role not only in image quality but also subsequently seed reconstruction accuracy.

When seeds are placed in close proximity of one another and a large pinhole is utilized this will ultimately result in a larger projection and may result in overlapping of two or more adjacent projections, inhibiting reconstruction capabilities. This phenomenon was present within the data and is shown in Figure 3. A comparison between the data from the 800 μm and 500 μm collimator including 15 seeds placed at varying heights from the pinholes (within a distance of approx. 5.5cm) was performed in order to determine the difference in projection sizes in the two data sets (Table 1).
It was concluded that the 800 μm collimator data showed a 22% increase in projection size when compared to the data of the 500 μm collimator. The large projection size results in projections overlapping neighbouring seeds, creating great uncertainty in the reconstruction of the centre of mass of the single seed. The overlapping of two seeds projections results in one single projection which is misinterpreted as a single seed with a centre of mass shifted. This leads to the large discrepancies recorded between CT and BracView-800 reconstructed final positions showed in Figure 2. Details of the reconstruction methods have previously been presented [9].

Table 1. 800 μm and 500 μm collimator projection sizes.

	800 μm Pixel width (mm)	Pixel height (mm)	500 μm Pixel width (mm)	Pixel height (mm)
	1.155	0.825	0.605	0.770
	0.935	0.880	0.605	0.605
	1.320	0.770	0.770	0.605
	1.100	0.715	0.770	0.660
	1.210	1.100	0.770	0.660
	1.045	0.990	0.715	0.495
	1.045	0.880	0.770	0.550
	0.990	0.770	0.825	0.550
	0.935	0.770	0.990	0.715
	1.375	0.825	0.880	0.550
	1.650	1.210	1.210	0.715
Average (mm) ± 0.006 mm	1.160	0.885	0.810	0.625

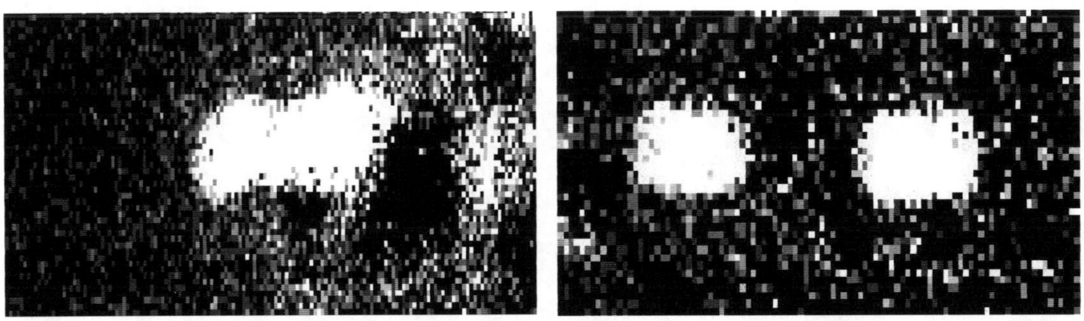

(a) (b)

Figure 3. a) Overlapping seed projections with use of the 800 μm collimator, inhibiting accurate delineation of individual seed center of mass for reconstruction, **b)** clear projection definition provided with use of the 500 μm collimator.

4. Conclusions

BrachyView data utilizing an 800 and 500 μm collimator pinhole has been analysed and compared in order to assess the effect on reconstruction capabilities of the system. Results from the 800 μm data showed increased baseline noise from scatter contribution and reconstructed positions with large inaccuracy. Only 70% of seeds have been reconstructed due to overlapping of seed projections which could not be used to identify the individual centre of masses. This is an indicator that the use of the 800 μm collimator for seed reconstruction before seed deployment from treatment needles is not feasible with the current methodology employed.

5. Acknowledgments

This research has been conducted with the support of the Australian Government Research Training Program Scholarship. The authors would like to thank the National Health and Medical Research Council for its support of the research and development of the BrachyView probe under award NHMRC Development Grant GNT1113631.

IOP Conf. Series: Journal of Physics: Conf. Series **1154** (2019) 012005 doi:10.1088/1742-6596/1154/1/012005

6. References

[1] Australian Government 2017 https://prostate-cancer.canceraustralia.gov.au/statistics
[2] Cancer Council Australia 2013 http://www.ncbi.nlm.nih.gov/pmc/articles/PMC3635047/
[3] Hurley C *et al* 2006 *Nucl. Instrum. Meth. A* **565** 801-11
[4] Pickles T *et al* 2010 *Int. J. Radiat. Oncol.* **76** 43-9
[5] Metcalfe P *et al* 2007 *The Physics of Radiotherapy X-rays and Electrons* (Madison, WI: Medical Physics Publishing Corp.)
[6] Dawson J E *et al* 1994 *Radiother. Oncol.* **32** 268-70
[7] Nag S *et al* 2000 *Int. J. Radiat. Oncol.* **46** 221-30
[8] Alnaghy S *et al* 2017 *Phys. Medica* **34** 55-64
[9] Petasecca M *et al* 2013 *Med. Phys.* **40** 041709

Characterization of a high spatiotemporal resolution monolithic silicon strip detector for MRI-linac dosimetry

T J Causer[1,2], T Chapman[1], B M Oborn[1,2], J A Davis[1], M Petasecca[1], A B Rosenfeld[1] and P Metcalfe[1]

[1]Centre for Medical Radiation Physics, University of Wollongong, NSW, Australia
[2]Illawarra Cancer Care Centre, Wollongong, NSW, Australia

E-mail: tjc721@uowmail.edu.au

Abstract. Multiple vendors are now offering real-time MRI-guided radiotherapy systems. Quality assurance of small fields delivered with an MR-RT system requires detectors with high spatiotemporal resolution, and magnetic insensitivity. High spatial resolution is required to characterise the asymmetric penumbra that transverse designs demonstrate. In this work the authors describe the characterisation of the dosimetric performance of a monolithic silicon strip detector mounted to a flexible polymide (Kapton) printed circuit board intended for use in MRI-linac dosimetry.

1. Introduction

There are now multiple vendors offering radiotherapy systems with online magnetic resonance (MR) image guidance. These systems can provide superior soft tissue contrast compared to kV photon image guidance technologies. The ViewRay MRIdian® Linac system combines 6MV flattening-filter free linac with a 0.35T MRI scanner. The Elekta Unity integrates a 7MV flattening-filter free linac system with a 1.5T MRI scanner [1-3]. In both systems the magnetic field axis is perpendicular to the beam axis. With this configuration the Lorentz force that acts on secondary electrons causes electron return effect (ERE) at high/low density interfaces and creates lateral asymmetry in beam penumbra [4].

To meet the needs of quality assurance of small fields delivered in the presence of a strong magnetic field detector systems with high spatial resolution, high temporal resolution and magnetic insensitivity are required. Monolithic silicon strip detectors have been shown to be suitable for small field dosimetry of linac beams [5]. The introduction of magnetic fields presents new challenges in detector design. In this work we characterise the dosimetric performance of a monolithic silicon strip detector mounted to a flexible polymide printed circuit board (PCB) intended for use in MRI-linac dosimetry.

2. Materials and methods

The monolithic silicon strip detector, (sDMG-256A) consists of 256 phosphorous implanted (n+) strips of size 20x2000μm^2 with 200μm pitch on a bulk p-type silicon substrate. The silicon strip array is wire bonded to a flexible polymide (Kapton) PCB carrier, shown in figure 1. The silicon and wire bonding are covered by a protective epoxy layer. The data acquisition (DAQ) system has been developed in house at the CMRP, it is based around the AFE0064 multichannel electrometer chip from Texas Instruments. To read out 256 channels, 2 analog-to-digital converters are used with 4

IOP Conf. Series: Journal of Physics: Conf. Series **1154** (2019) 012006 doi:10.1088/1742-6596/1154/1/012006

AFE0064's all synchronised by an FPGA. The data is passed from the FPGA to the user's computer via USB2.0 and accessed with a graphical user interface [6].

Dosmetric characterization of sDMG-256A included dose linearity investigated over the range (20-1000)cGy. Dose per pulse dependence (DPP) was measured over the range (0.29-4.66) x 10^{-4} Gy/pulse. The set up was a 10x10cm^2 field with the detector placed a 5cm depth in solid water, 10cm backscatter and the SSD varied from (70-300) cm. At each measurement point a CC13 ionisation chamber reference measurement was taken and corrected for recombination using two voltage method.

The angular dependence was investigated using the Dosepoint RT-smartIMRT® phantom. The detector was setup in the phantom such that the central detector channel was coincident with the central axis of rotation of the Dosepoint phantom. The linac gantry was stationary and the Dosepoint phantom was rotated, measurements taken every 15°.

Uniformity of detector channel response was accessed using a 6MV, 20x20cm^2 field, with the detector at depth 10cm in a solid water phantom positioned at 100cm SSD. The uniformity is calculated as the differential response of each channel relative to the central channel. Measurements were performed on a Clinac 2100EX, Varian Medical Systems using 6MV and 10MV beams.

The effect of air gap width in detector packaging for small field output factor (OF) measurements in the presence of 1.2T transverse magnetic field was investigated. The magnetic field is generated using a portable permanent magnet system shown in figure 1. A bespoke phantom was required to position the detector array in the gap between the cones of the permanent magnet and to allow precise changes to the airgap above the diodes. Measurements are performed at an extended SSD as the magnetic apparatus does not physically fit closer to the linac source, this extended distance also allows the fringe field of the magnet to drop below expected background levels at the linac head as not to affect the linac operation. Measurements were taken at 150cm SAD for field sizes 3, 5, 10, 15mm^2 and air gaps 0, 0.36, 0.72, 1.08mm. A replica of the magnet system that does not contain the neodymium-iron-boron magnets was used to perform zero field measurements with the same scatter conditions, in order to isolate magnetic field effects on the detectors response.

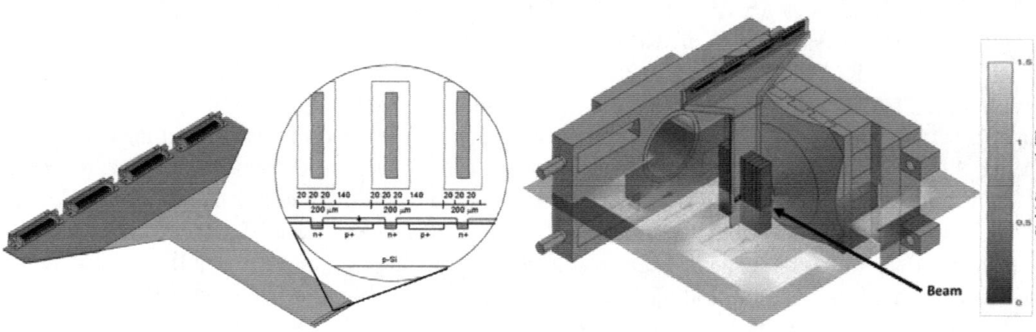

Figure 1. Left - sDMG-256A, green is FR4 section of PCB, orange is the flexible polymide. Right Output factor measurement setup with portable permanent magnet system.

3. Results

The dose response of sDMG-256A for (20-1000) cGy shown in figure 2 displays a linear response (R^2 = 1). The dose per pulse dependence shown in figure 2 displays a 10.3% variation over the range (0.29-4.65) x 10^{-4} Gy/pulse. The uniformity of the sDMG-256A in figure 3 shows before applying the equalization vector detectors channel response relative to the central channel for 99% of channels were within ±5%. After channel normalisation the variation of all channels was within 0.2%. The angular response of the sDMG-256A's central channel is displayed in figure 3. A 20% decrease in response was observed for beam angles parallel to the long axis of the detectors monolithic silicon chip. The effect of an air gap above the detectors sensitive volume is known to decrease the OF of diodes and has been used to correct for the diodes overresponse to small fields [7]. Comparing the OF for a range of air gaps and field sizes it is observed as shown in figure 4 that an increase in airgap width results in a decreased in measured OF in a transverse magnetic field with the effect being larger for smaller field sizes compared to the B=0 case. This implies that the method of compensating the overresponse of diodes to small field OF's cannot be translated into an MRI-linac environment. This decreased response in the presence of a transverse magnetic field is due to ERE of secondary electrons within the air gap above the silicon diodes.

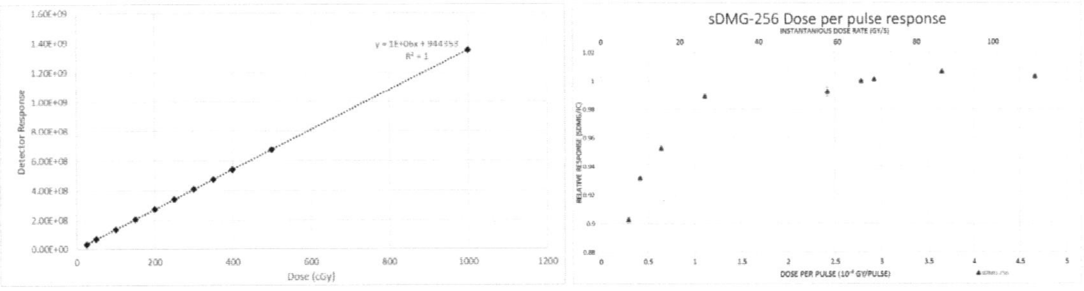

Figure 2. Left – dose linearity, Right – sDMG-Dose per pulse response normalized to 2.78x10^{-4} Gy/pulse.

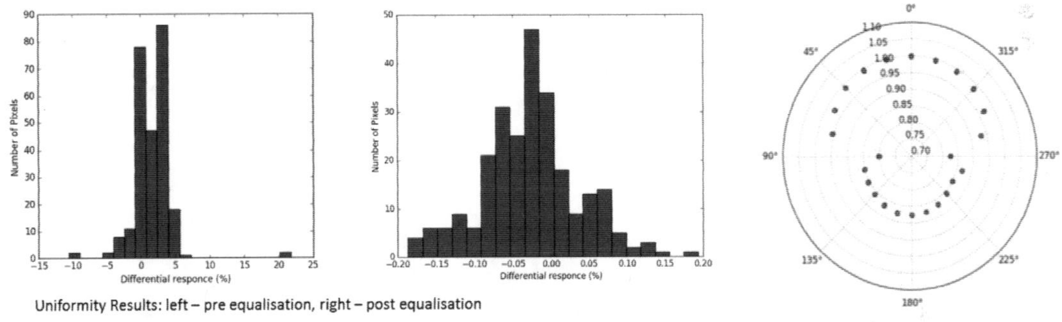

Figure 3. Left - histogram of uniformity results pre-equalisation, Centre - histogram of uniformity results post equalisation, Right - Angular response of central channel.

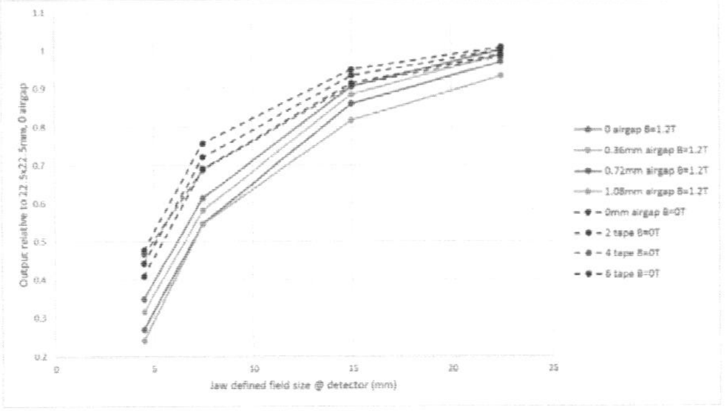

Figure 4. Effect of air gap above detector on small field output factor measurement (solid lines – B=1.2T, dashed lines – B=0T).

4. Conclusion

The preliminary dosimetric characterisation of a monolithic silicon strip detector mounted to flexible polymide carrier has been carried out. The sDMG-256A has been tested at 1.2T with small fields. Preliminary results describe a complex change in output factor that depend on B-field and air-gap. We are working on an insert that will ameliorate the effect of the air gap above the dosimeter. Further

IOP Conf. Series: Journal of Physics: Conf. Series **1154** (2019) 012006 doi:10.1088/1742-6596/1154/1/012006

work will be required to describe a complete characterisation of the response of the system for various magnetic fields and x-ray beam field sizes.

5. References

[1] Gargett M *et al* 2015 *Med. Phys.* **42** 856-65
[2] Mutic S *et al* 2014 *Semin. Radiat. Oncol.* **24** 196-9
[3] Raaymakers B W *et al* 2017 *Phys. Med. Biol.* **62** L41
[4] Raaymakers B W *et al* 2004 *Phys. Med. Biol.* **49** 4109-18
[5] Aldosari A H *et al* 2014 *Med. Phys.* **41** 91707
[6] Fuduli I *et al* 2014 *J. Instrum.* **9** T06003
[7] Al Shukaili K *et al* 2017 *Med. Phys.* **44** 628-36

IOP Conf. Series: Journal of Physics: Conf. Series **1154** (2019) 012007 doi:10.1088/1742-6596/1154/1/012007

Evolution of Diamond based Microdosimetry

J A Davis, M Petasecca, S Guatelli, M L F Lerch and A B Rosenfeld
Centre for Medical Radiation Physics, University of Wollongong, NSW 2500, Australia

E-mail: jeremyd@uow.edu.au

Abstract. The requirements for solid state micro- dosimetry particularly within the context of medical and space environments necessitate a tissue equivalent and radiation hard material, for which diamond is uniquely suited. An overview of the current status of diamond based microdosimetry is given. This overview will explore previous and current technologies developed by the Centre for Medical Radiation Physics (CMRP). The overview will analyse technologies in terms of their advantages and disadvantages within the context of microdosimetry. Using this as a basis, recommendations with respect to fabri- cation methodologies/techniques are provided in order to direct the course and progress of development so as to achieve true diamond based microdosimetry.

1. Introduction

Diamond based radiation detector technologies is a relatively new area of study in the field of solid state microdosimetry. This interest in diamond is attributed to its radiation tolerance [1, 2] and tissue equivalence [3-7]. Diamond based microdosimeters whilst still under development, are under consideration for application in hadron therapy and radioprotection applications in harsh radiation environments where their radiation hardness will enable long term operation.

Microdosimetry is a methodology which considers stochastic radiation interactions and associated biological effects for Sensitive Volumes (SV) on the scale of the cell. Key to microdosimetry is the concept of lineal energy (y) which is defined as the quotient of an energy deposition event (s) and the mean chord length of the sensitive volume (l) [8]. Microdosimetric Spectra are typically plotted in a semi-logarithmic scale as events span orders of magnitude. The two most commonly used spectra are frequency ($yf(y)$ vs. $log(y)$) and dose ($y^2 f(y)$ vs. $log(y)$) distributions, where equal areas under the curve represent equal fractions of events and fractional doses respectively. The experimental determination of these quantities then requires detectors/microdosimeters with accurately defined micron sized SVs and thus a well-known mean chord length (l).

At the Centre for Medical Radiation Physics (CMRP), University of Wollongong, a variety of different designs and fabrication processes have been utilised in the attempt to realise an array of well-defined and accurately defined micron scaled SVs. In this paper, an overview of the current status of diamond based microdosimetry at CMRP is given. In addition, an examination of the most promising fabrication technologies currently available that might be applicable to diamond microdosimetry will be examined. This work will feature the various technologies currently available and will assess each with respect to the fabrication techniques utilised in their creation within the context of their ability to satisfy the aforementioned solid state microdosimeter requirements.

2. Fabrication techniques for diamond microdosimetry

Whilst there is extensive literature available regarding the fabrication of diamond-based radiation detectors, the area of diamond microdosimeters remains relatively untouched. This section will detail the positive and negative implications of the more promising techniques currently available.

2.1 Ion implantation

Electrically isolated SV structures may be created by means of ion implantation. This technique involves the controlled bombardment of energetic charged ions to create a buried electrode structure. The depth of structure is determined by the ion energy as defined by the Bragg peak. The first use of this technique for the purpose of microdosimetry is outline by Davis *et al* 2012 [9] in the creation of a metal-insulator-p-type diamond (M-I-P) structure. Ion implantation of 2 MeV boron through a patterned aluminium mask was utilised to create a 3D wall-less SV structure at approximately 1.38 μm depth in Type IIa electronic grade single crystal diamond from Element Six. The fabrication methodology employed in the creation of this device was successfully used to create an electrically isolated array of SVs with scalability to achieved SVs appropriate for microdosimetry, albeit with very low charge collection efficiency. This under-performance of the device with respect to CCE has been hypothesised to be the result of radiation induced damage via the boron implantation technique. The normal damage process due to ion implantation in diamond is the creation of vacancies and interstitials, which may act as charge traps and scattering sites. Annealing is generally sufficient to remove such damage and electrically activate the implanted impurities. The lower than expected CCE indicates that either the implantation or annealing methodology requires further investigation in order to be useful for diamond-based devices.

2.2 Laser ablation

Physically isolated volumes may be created via laser ablation as opposed to the electrical isolation achieved via ion implantation. This technique was utilised to create a conceptually simple planar metal-insulator-metal (MIM) structure utilising high purity single crystal chemical vapour deposited (scCVD) diamond sourced from Element6 [10]. Laser ablation is utilised to create 'trenches' surrounding the planar Metal-Insulator-Metals structure in order to achieve SV definition. The benefits of this technique is that SV definition may be defined post-growth, through lithographic and laser milling processes. It should be noted that whilst SV may be reduced in two dimensions, there is still a dependence upon the thickness of the sample due to planar structure of the device. The primary advantages of laser ablation over alternative techniques such as deep reactive ion etching, is the rate and ease at which diamond may be micro-machined to a desired depths and patterning. This device demonstrated good charge confinement and true promise in terms of applicability as a diamond based microdosimeter. The devices have demonstrated charge signal amplitudes matching the theoretical value, i.e., approximately 100% CCE. The IBIC experiments showed that the technique of pulsed laser milling is capable of creating physically isolated SVs, leading to reduced effect of charge sharing between adjacent pixels. Charge collection confinement was shown to be affected by particle energy and angle of incidence, i.e., penetration depth. In addition, it was shown that the charge collection confinement could be further improved/optimized though appropriate adjustment of the electric field strength. Lastly, it was shown that a combination of the fabricated laser milled trenches and surrounding electric fields when the device is in operation mode results in a well-defined and laterally confined charge collection region. However, the planar MIM structure means that there is an inherent dependence upon the sample thickness.

2.3 Active Brazing Alloy

The use of lateral electrode type structures in diamond detectors is a relatively new area of re- search. One of the more prominent methods has been in the utilisation of fempto second pulsed lasers to create graphitic columns/pillars within the diamond bulk, and whilst ideal in terms of tissue equivalence, leads to many issues in charge collection properties [11]. One alternative is the uses of

IOP Conf. Series: Journal of Physics: Conf. Series **1154** (2019) 012007 doi:10.1088/1742-6596/1154/1/012007

commercially available Active Brazing Alloys (ABA). These materials commonly gold or silver based contain an active ingredient (1-5%), to pro- mote reaction and wetting/adhesion with 'ceramic' surfaces. For applications involving diamond, ABAs containing titanium are the active ingredient of choice given their closely matched work function (φ=4.33 eV) and this ability to form a carbide. ABAs are a versatile material, able to coat surfaces and/or fill voids. The use of ABAs in combination with patterned laser ablation provides a simple and effective means for creating lateral electrodes within diamond. The performance of Silver ABA for metallisation of diamond for radiation detection applications was shown by Davis *et al* 2016 [12]. In this paper, silver ABA was used to fill laser milled regions to create a 3D lateral electrode structure (3D-LES). Unlike the two previous prototypes, this is the first design to employ a lateral electric field structure. This device achieves 100% CCE in the region of interest and along with laser ablated boundary trenches is able to successfully isolate charge collection from surrounding regions [12]. Whilst the 3D-LES detector is not yet ready for microdosimetry, it does appear to satisfy the requirements of a radiation detector for high energy physics experiments, i.e. radiation hardness, fast detection and large charge collection distances and efficiency. With the appropriate adaptations of the designs, ABA technologies is suitable for the creation of microdosimetric technologies. Although it should be noted that applications within photon dosimetry where tissue equivalence is paramount, the use of high Z metals like silver and titanium would not be desirable.

2.4 Multi-layered diamond

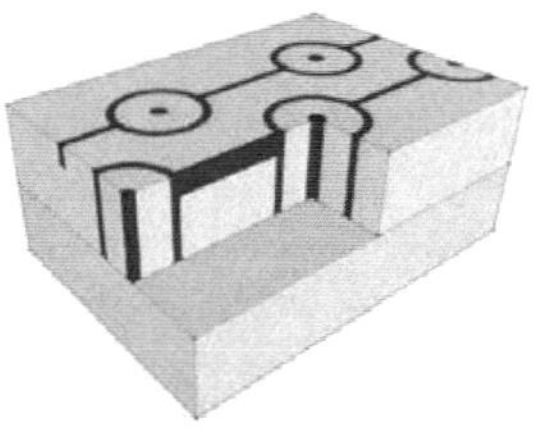

Figure 1. Cross sectioned schematic of the 3D-LES-2 microdosimeter prototype constructed within a bi-layered pcCVD diamond structure. The electronic grade device layer is depicted in white whilst the low quality substrate is depicted in yellow. The silver ABA filled trenches created by laser ablation used to create the inner/outer electrode and connecting bridge structures are depicted in grey.

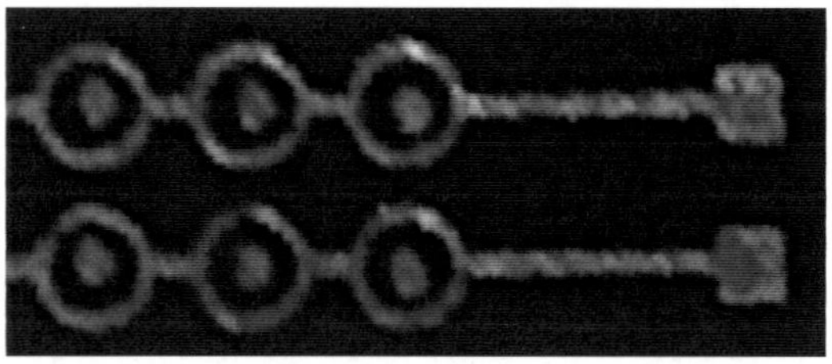

Figure 2. Microscope image of the 3D-LES-2 prototype. The grey concentric circles connected by a thin strip are the outer electrodes. Each row is connected to common pad for electrical connection. At the centre of each outer electrode is the silver ABA filled inner electrode. The diamond here appears black given the optical transparent nature of the device layer.

The common issue with previous designs dis- cussed, is the dependence upon the diamond sample thickness, resulting in SV structures too large to be applicable for microdosimetry. Whilst it is possible to purchase thin diamond (10-30μm), the samples are extremely fragile and liable to fracture even under the small amounts of force utilised in contact-based photolithography. To that end, CMRP is instead investing its efforts in the development of bi/tri-layered diamond structures, such as has already been developed by the University of Rome Tor Vergata laboratories [13]. The bi-layered structure requires the growth nominally intrinsic (Type IIa) diamond upon low grade commercial single crystal High Pressure High Temperature (HPHT) diamond. The tri-layered structure utilises a two-step diamond deposition process upon low grade commercial single crystal High Pressure High Temperature (HPHT) diamond. The first deposition phase involves the growth of single crystal Boron Doped Diamond (BDD) upon the HPHT substrate. The second phase is the growth of nominally intrinsic (Type IIa) diamond, upon the layer of BDD. The thickness of each layer can be customised, given a known growth rate. The end result is a layer of high quality diamond with customisable thickness.

The use of layered diamond growth processes in combination with suitable post processing (Laser ablation and/or Dry Reactive Ion Etching etc) provides the most reliable means or creating a pixelated diamond microdosimeter device. To that end, CMRP has produced a prototype diamond micro- dosimeter utilising a combination of laser ablation and active brazing alloys upon a bi-layered structure as a proof of concept for next generation diamond microdosimetry. The 3D-LES-2 structure features cylindrical SV structures created by laser ablation and ABA situated within a thin membrane of electronic grade pcCVD diamond grown at the Melbourne Centre for Nanofabrication (MCN). Patterned laser ablation is used to create an array of concentric circles to form the outer electrode structure with further laser ablation used to connect rows of outer electrode in parallel. A final round of laser ablation is used to mill hollow cylinders at the centre of each of the aforementioned concentric circles, followed by the application of silver ABA to replace ablated diamond and create electrical contacts. Currently the minimum size of the inner electrode is restricted such that the diameter is greater or equal to 50μm to allow for direct wire bonding. In the case of the bi-layered device, the depth of laser ablation was performed such that both the inner and outer electrode penetrated into the low-quality diamond substrate. To better illustrate the concept of the 3D-LES-2 device, a conceptual schematic is depicted in Fig. 1. Electrical characterisation of the device has now been completed with charge collection characterisation to be performed in the near future. The results of this study will be detailed in an upcoming paper.

3. Discussion and Conclusion

Diamond is a highly interesting material for radiation detectors in a variety of applications from high energy physics to medical physics. CMRP is at the forefront of 3D pixelated detectors for the purpose of microdosimetry. Diamond based radiation detectors can be developed utilising a variety of different fabrication techniques. Previous device structures have had SV size dependent upon the thickness of the diamond sample or a loss in CCE in the case of ion implantation. The future of diamond based microdosimetry will utilise multi- stage diamond growth to produce a layered structure which in combination with the techniques discussed in this paper will allow for true microdosimetric sized SVs in diamond to be realised.

4. Acknowledgements

The authors wish to acknowledge the existing collaboration with the University of Melbourne in the joint fabrication of diamond detectors. This project was partially supported by the Australian Research Council (ARC) Grant No. DP 1096600.

5. References

[1] Bauer C *et al* 1995 *Nucl. Instr. Meth. Phys. Res. A* **367** 207- 11
[2] Sato Y *et al* 2015 *Nucl. Instr. Meth. Phys. Res. A* **784** 147-50
[3] Planskoy B 1980 *Phys. Med. Biol.* **25** 519-32
[4] Yacootts A *et al* 1990 *Phys. Med. Biol.* **35** 1409-22
[5] Heydarian M *et al* 1993 *Phys. Med. Biol.* **38** 1035-42
[6] Davis J A *et al* 2014 *IEEE Transactions on Nuclear Science* 61 1544-51
[7] Gorjiara T *et al* 2011 *Phys. Med. Biol.* **56** 4685-99
[8] Rossi H and Zaider M 2012 *IEEE Transactions on Nuclear Science* **59** 3110-6
[9] Davis J A *et al* 2012 *IEEE Transactions on Nuclear Science* **59** 3110-16
[10] Davis J A *et al* 2014 *IEEE Transactions on Nuclear Science* **61** 3479-84
[11] Caylar B *et al* 2013 *Appl. Phys. Lett.* **103** 7-10
[12] Davis J A *et al* 2017 *Appl. Phys. Lett.* **110** 013503
[13] Marinelli M *et al* 2001 *J. Appl. Phys.* **89** 1430-5

IOP Conf. Series: Journal of Physics: Conf. Series **1154** (2019) 012008 doi:10.1088/1742-6596/1154/1/012008

BNCT dosimetry: peculiarities and methods

G Gambarini[1,2,3], D Bettega[2,3], A Gebbia[2], E Artuso[2], M Felisi[2], D Giove[2,3], V Klupak[4], L Viererbl[4] and M Vins[4]
[1]Università degli Studi di Milano, Department of Physics, Milano, Italy.
[2]INFN Istituto Nazionale di Fisica Nucleare, Section of Milan, Milano, Italy
[3]Department of Neutron Physics, Research Centre Řež, Czech Republic

E-mail: grazia.gambarini@mi.infn.it

Abstract. Dosimetry in tissue exposed to the epithermal neutron beams utilized for BNCT is complex, due to the multiplicity of the possible neutron reactions and consequently of the secondary radiation that contains photons, charged particles and recoil nuclei. Owing to the different radiobiological effectiveness of the various components of the absorbed dose, it is necessary to attain the evaluation of each of them. In addition, the spatial distributions of these dose components changes considerably with size and shape of the irradiated volume. Therefore, BNCT dosimetry requires suitably developed calculations and experimental methods. In this work, Monte Carlo simulations in phantoms of different sizes and shapes have been developed. Experimental methods for separating the dose components, mainly based on gel dosimeters and thermoluminescence detectors, have been applied. Moreover, the change in the absorbed dose resulting from the addition of ^{157}Gd was investigated. Both measurements and calculations have been done with the BNCT epithermal beam of the LVR-15 reactor.

1. Introduction

BNCT dosimetry is a challenging task, particularly in the case of irradiations with epithermal neutrons, as required for deep tumor treatments. In tissue, the epithermal neutrons are moderated mainly by the elastic scattering interactions with hydrogen nuclei ^{1}H(n,n')^{1}H and the energy spectrum of BNCT neutron beams is suitably designed in order to get high fluence of thermal neutrons in a volume centered at a depth of about 2 cm. The radiotherapy dose is due to the charged particles generated in the reactions of thermal neutrons with ^{10}B selectively accumulated in the tumor tissue (^{10}B(n,α)^{7}Li), here referred to as D_B. In addition to D_B, it is necessary to take into account the dose from all the secondary radiation generated by neutrons, both in tumor and in healthy tissue. The dose contributions that have to be considered can be divided into three general groups, each characterized by a different spatial distribution: the dose coming from the charged particles generated by thermal neutron reactions with particular isotopes, the gamma dose (D_γ) and the dose due to epithermal- and fast-neutron elastic and anelastic collisions (D_{fast}).

The charged particles have a short range and then a spatial distribution directly related to the distribution of the thermal neutron fluence Φ_{th}, and the absorbed dose can be evaluated, in each position, by means of the corresponding kerma factor. The spatial distribution of Φ_{th} is considerably dependent on both size and shape of the irradiated volume. A significant percentage of the absorbed dose comes from gamma radiation, partially due to background (not considered in this work) and partially to the photons of 2.2 MeV emitted in the reactions of thermal neutrons with hydrogen (^{1}H(n,γ)^{2}H). This emission, in each point of the irradiated volume, is linearly correlated to the thermal

IOP Conf. Series: Journal of Physics: Conf. Series **1154** (2019) 012008 doi:10.1088/1742-6596/1154/1/012008

neutron fluence in that point but, owing to the long range in tissue of the 2.2 MeV photons, the resulting dose (D_γ) undergoes changes, due to variations of shape or size of the irradiated volume, greater than those of the thermal neutron fluence.

A lower but significant dose contribution is given by neutrons with E > 0.5 eV, whose fluence is here referred to as Φ_{fast}. The spatial distribution of D_{fast} is not greatly dependent on the geometry of the irradiated volume.

Both D_B and D_γ have spatial distributions significantly dependent on size and shape of the irradiated volume and on the direction of incidence of the neutron beam. In order to estimate the influence of the unavoidable lack of precision of the input data for dose measurements or calculations, for the issues described above, appropriate experimental and computational evaluations have been carried out.

Moreover, a few evaluations on neutron fluence and gamma dose in phantom containing ^{157}Gd have been done, because this isotope has been proposed in place of or in addition to ^{10}B, given the effectiveness of Auger electrons in destroying tumor cells. Owing to the high cross section (255000 b) of the reaction $^{157}Gd(n,\gamma)^{158}Gd$, a considerable reduction of Φ_{th} and a noticeable increase of D_γ are expected.

2. Materials and methods

Experimental measurements for BNCT dosimetry require suitably developed methods, because: (i) it is necessary to separate the various dose components owing to their different radiobiological effectiveness; (ii) it is necessary to determine the spatial distribution of each dose component; (iii) the determinations have to be made for the specific geometry of the irradiated volume and (iv) most standard dosimetry methodologies are not applicable with the high neutron fluence-rate proper of BNCT beams (about 10^9 $cm^{-2}\,s^{-1}$). A type of detection modality that has shown to be very advantageous, in particular in the case of BNCT epithermal neutron beams, is the gel dosimetry [1]. In fact, gel dosimeters [2-9] allow to determine the spatial distribution of the absorbed dose and, by means of suitable variation of the isotopic composition, they give the possibility of separating all the various dose contributions [10]. If gel dosimeters are prepared in large volumes, the precision of the dose distributions is scarce, because a change of the isotopic composition causes a change of neutron transport. On the opposite, if gel dosimeters are prepared in layers 3 mm thick or similar, inserted in tissue-equivalent phantoms, the accuracy of the determined dose distributions has proven to be high. In phantoms, dose mapping can be attained by means of thermoluminescence detectors (TLDs), with which it is possible to determine D_γ and Φ_{th}. Activation foils can also be used for Φ_{th} measurements, but very small foils and cadmium covers are necessary to avoid sensible change of neutron transport. Concerning the fast neutron dose, a method based on twin ionization chambers has been proposed.

Monte Carlo calculations are widely developed for BNCT dosimetry, based on phantoms of various shapes: cylindrical, ellipsoidal or anthropomorphic. However, a specific determination would be necessary for each specific case of treatment.

In this work, some results are reported, obtained with Fricke-XO gel dosimeters and TLDs. The standard gel dosimeters are water-equivalent and then the absorbed dose, after exposure to epithermal neutrons, is due to photons and to fast neutrons, mainly recoil-proton dose. If a suitable amount of ^{10}B is added to the gel composition, also the charged particles generated by ^{10}B reactions give a contribution to the absorbed dose. If standard gel dosimeters are prepared with heavy water instead of water, fast neutron contribution comes through recoil-deuterons. With suitably developed methods, images and profiles of the various dose contributions are obtained.

Moreover, many MC calculations concerning neutron spectrum and neutron fluence profiles have been performed, in order to investigate the possible variations due to changes of size or shape of the irradiated volume [11]. The last results are here reported.

IOP Conf. Series: Journal of Physics: Conf. Series **1154** (2019) 012008 doi:10.1088/1742-6596/1154/1/012008

3. Results

Several measurements and calculations have been performed with the epithermal neutron beam of the LVR-15 research reactor, designed for BNCT at the Research Centre Řež. Different water phantoms have been experimented: standard (40x40x20 cm^3), cubic (20x20x20 cm^3) and spherical (diameter 16 cm). MC simulations have been developed to obtain energy spectra, longitudinal or transversal profiles of D_γ, Φ_{th} and Φ_{fast}. Experimental images of D_γ, D_{fast}, D_B and Φ_{th} have been obtained in the standard phantom and in a cylindrical one (14 cm in height, 16 cm in diameter). Considerable variations of Φ_{th} and even more of D_γ were found in the various different phantoms. In Figure 1, some results of MC calculations are reported.

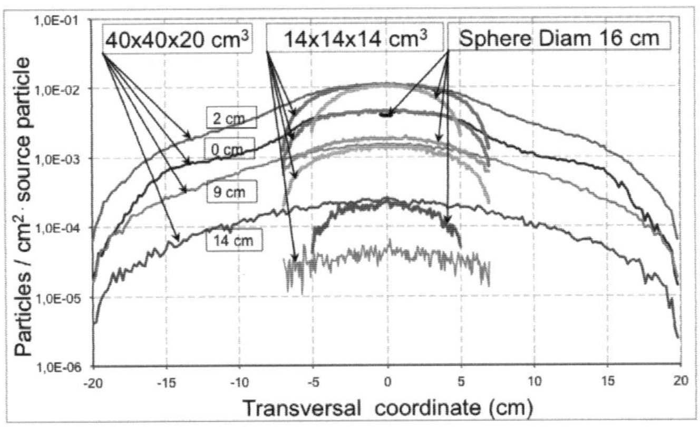

Figure 1. Calculated transversal profiles of thermal neutron fluence in the standard (40x40x20 cm^3), cubic (14x14x14 cm^3) and spherical (Ø = 16 cm) phantoms, at depth of 0 cm, 2 cm, 9 cm and 14 cm.

In Figure 2, the calculated profile of Φ_{fast} and the measured profile of D_{fast} are shown and in Figure 3 the ratio D_{fast}/Φ_{fast} is reported. In order to investigate the cause of this difference of the trends vs depth, some considerations about the change of neutron energy-spectrum with increasing depth have been done. In Figure 4, the energy spectra from 0.5 eV onwards, in water at different depths, are reported. The integral of the fluence values in three regions of energy has been evaluated: 10^{-5}-10^{-3} MeV, 10^{-3}-10^{-1} MeV and 10^{-1}-10 MeV. In Figure 5, the results for the various depths are reported. It is evident that, with increasing depth, the relative contribution of the lower energies decreases and that of the higher energies increases. This result is consistent with the increase of the ratio D_{fast}/Φ_{fast} with increasing the depth in water.

Some results showing the great increase of D_γ in presence of 100 μg/g of ^{157}Gd are reported in Figure 6 (Monte Carlo) and Figure 7 (experimental). The experimental results were attained with gel dosimeters (standard and containing ^{157}Gd) by means of suitable algorithms and with TLDs.

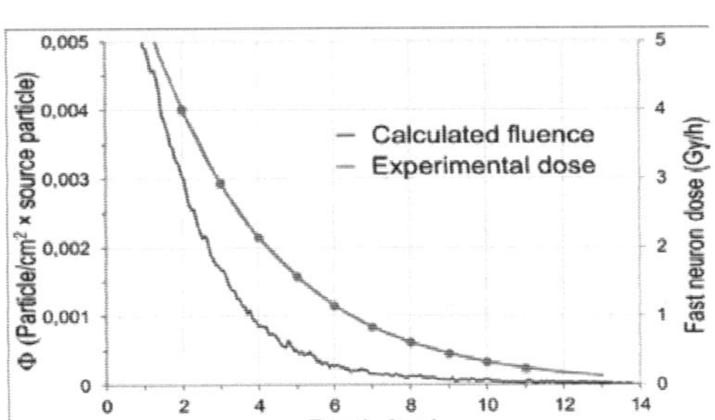

Figure 2. Profiles, along the beam axis, of D_{fast} and of the fluence of neutrons with E > 0.5 eV.

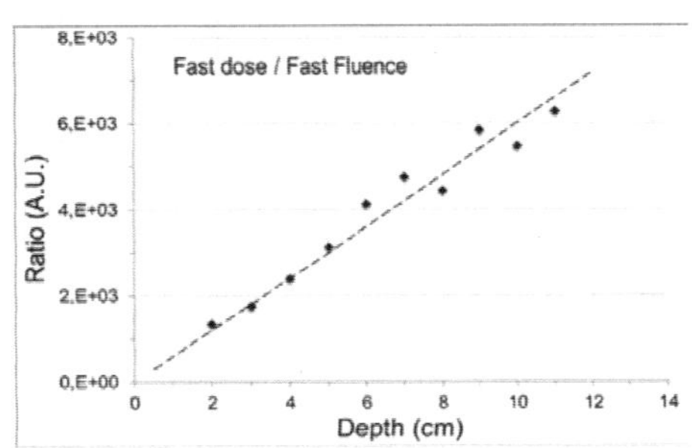

Figure 3. Ratio between the dose and fluence values reported in Figure 2.

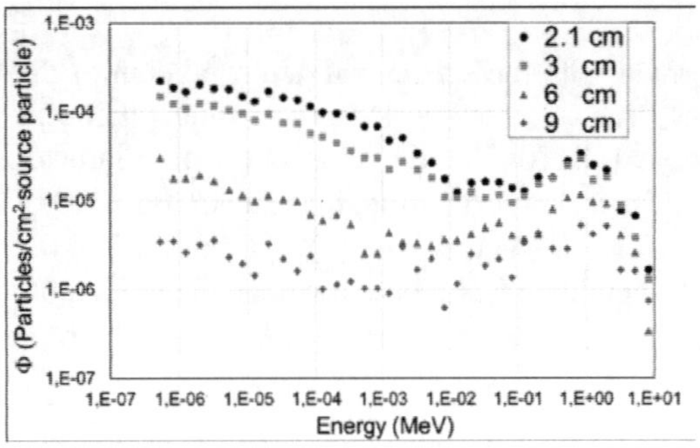

Figure 4. Energy spectra of neutrons with E > 0.5 eV at depths in water of 2.1, 3, 6 and 9 cm.

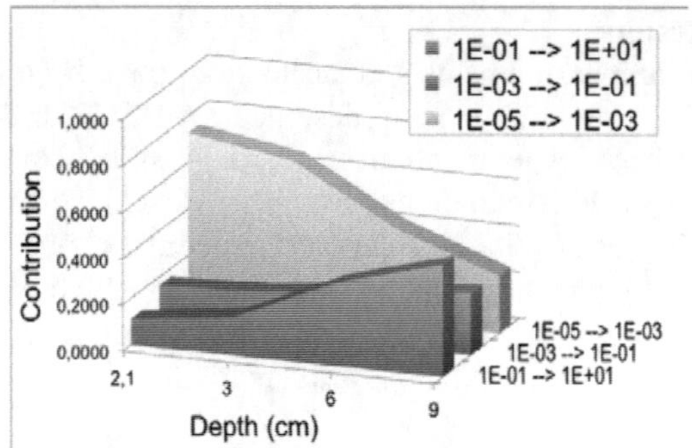

Figure 5. Integral neutron fluence in three regions of energy, vs depth in water.

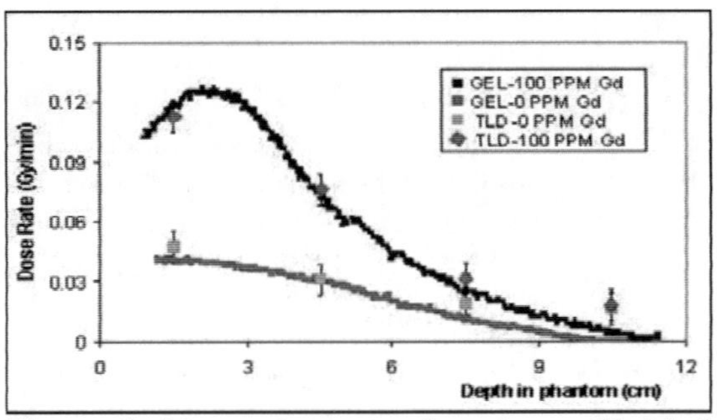

Figure 6. Measured gamma dose in a cylindrical phantom of water containing *100* μg/g of ^{157}Gd and of water without ^{157}Gd.

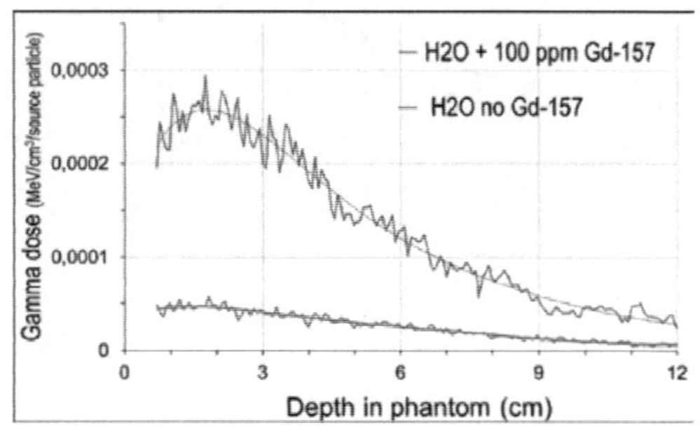

Figure 7. Calculated gamma dose in a cubic phantom of water containing *100* μg/g of ^{157}Gd and of water without ^{157}Gd.

4. Conclusions

The outcomes obtained with the various detectors were mutually consistent and complementary. The results of the work can give useful information on the topic. In particular, they allow a first estimate of the extent of the changes that may occur by changing the shape or size of the irradiated volumes.

5. References

[1] Khajeali A *et al* 2015 *Appl. Radiat. Isot.* **103** 72
[2] Hill B *et al* 2002 *Phys. Med. Biol.* **47** 4233-46
[3] Healy B J *et al* 2003 *Med. Phys.* **30** 2282-91
[4] Mather M L *et al* 2003 *Phys. Med. Biol.* **48** N269-75
[5] Hill B *et al* 2005 *Med. Phys.* **32** 1589-97
[6] Venning A *et al* 2005 *Nucl. Instrum. Meth. A* **555** 396-402
[7] Hurley C *et al* 2006 *Nucl. Instrum. Meth. A* **565** 801-11
[8] Baldock C 2009 *J. Phys.: Conf. Ser.* **164** 012002
[9] Gorjiara T *et al* 2011 *Phys. Med. Biol.* **56** 4685-99
[10] Gambarini G *et al* 2015 *Radiat. Phys. Chem.* **116** 21
[11] Gambarini G *et al* ASTM, STP160 (in press)

IOP Conf. Series: Journal of Physics: Conf. Series **1154** (2019) 012009 doi:10.1088/1742-6596/1154/1/012009

The Liver INSPECTR Trial: Towards improved understanding of liver function following radiotherapy

N Hardcastle[1, 2], C Haddad[3], G Schembri[4], D Bailey[4], B Jones[5], A Briggs[3], G Hruby[3] and A Kneebone[3]

[1]Department of Physical Sciences, Peter MacCallum Cancer, VIC, Australia
[2]Centre for Medical Radiation Physics, University of Wollongong, NSW, Australia
[3]Northern Sydney Cancer Centre, Royal North Shore Hospital, NSW, Australia
[4]Department of Nuclear Medicine, Royal North Shore Hospital, NSW, Australia
[5]Department of Hepatology, Royal North Shore Hospital, NSW, Australia

E-mail: nick.hardcastle@petermac.org

Abstract. Liver stereotactic ablative body radiotherapy (SABR) is an emerging treatment option for primary liver cancer and liver metastases. Liver SABR can provide high rates of local control however tumour doses are limited by dose to surrounding organs at risk including liver and gastrointestinal structures. Dose limits to liver are based on anatomical liver, without understanding of underlying liver function. This report describes a prospective clinical trial which measures the spatial variance of liver function through Tc-99m Mebrofenin SPECT in the context of liver SABR.

1. Introduction

There is an emerging body of evidence suggesting stereotactic ablative body radiotherapy (SABR) is an effective localised treatment modality for primary cancers of the liver and liver metastases. Recent reports suggest SABR results in comparable local control and quality of life outcomes compared with other localised treatment modalities [1]. SABR involves the delivery of a highly conformal high dose of radiotherapy in a small number of treatment sessions, typically 3-6. Dose is minimised to healthy liver to reduce risk of radiation induced liver disease (RILD). Frequency of RILD is low 1-5% with modern constraints however, more relevant is deterioration in liver function, for example by increasing Child-Pugh score by 2 or more which can be in the order of 30% [2]. Liver function end points and their dependence on radiation dose have been measured previously using global liver function tests. Mean liver dose has been linked to liver function decline, as well as dose to 700 ml of liver [3]. These liver function dose-responses however have been derived from anatomical liver [3]. Patients however often have multiple co-morbidities that render sections of the liver non-functional, thus dose-response data is clouded by the fact that the radiotherapy dose may have been delivered to non-functional liver [3].

Tc-99m mebrofenin SPECT imaging procedure allows evaluation of global and local liver function [4]. Mebrofenin, bound to albumin, is taken up by hepatocytes in a similar manner to organic anions such as bilirubin [5]. SPECT imaging with corresponding CT allows visualisation of the local liver function with respect to anatomy [5-8]. Measurement of regional liver function in the context of liver SABR has the potential to further refine dose-volume predictors of RILD and provide insight into liver function deterioration and recovery post-treatment. Moreover, if liver function is known on a local basis we may be able to spare highly functional liver more than we do currently through advanced optimisation and delivery methods such as volumetric modulated arc therapy (VMAT) [9].

This work describes the results obtained from the first four patients enrolled in a prospective observational trial investigating the role of Tc-99m Mebrofenin SPECT in liver SABR for both measurement of pre-SABR liver function and change of liver function as a result of SABR.

2. Materials and Methods

Liver function Investigation with Single Photon Emission Computed Tomography after Radiotherapy (Liver INSPECTR) is a Phase I pilot study investigating Tc-99m Mebrofenin SPECT in the context of liver SABR. The primary endpoint of the study is to evaluate the pattern of local and global liver function as demonstrated on Tc99m mebrofenin SPECT before and after liver SABR. Patients who are to receive SABR for treatment of primary liver cancer or liver metastases are eligible for the study. Participants receive a Tc-99m Mebrofenin SPECT-CT image at time of CT simulation, and at 1 and 6 months post SABR treatment.

SABR treatment planning 4DCTs were acquired and the treatment plan was performed on the average of the 4DCT, using an internal target volume (ITV) to account for tumour motion due to respiration. A 5 mm ITV-PTV isotropic margin was applied. Treatment planning was performed using VMAT, with an isotoxic prescription regime used; 50 Gy in 5 fractions is prescribed initially. If liver dose constraints were not met, the dose was reduced in 2.5 Gy increments until the liver dose constraints were met. The Tc-99m Mebrofenin SPECT-CT scans were acquired with a low dose 3DCT followed by acquisition of dynamic planar scintigraphy with SPECT on a Siemens Symbia SPECT-CT. For SPECT imaging the radiotherapy treatment planning position was replicated as close as possible to improve spatial correlation between SPECT-CT and radiotherapy treatment planning CT geometry.

The CT component of the SPECT-CT at each time point was rigidly registered to the treatment planning CT based on a best match of anatomical landmarks within the liver and the liver border in the vicinity of the tumour. The same rigid registration was applied to the SPECT image data, to result in SPECT data in the same frame of reference as the treatment planning CT, structures and dose. For each SPECT image, the total SPECT counts in non-tumour liver in each radiotherapy isodose bin from 0-50 Gy in 5-10 Gy increments was measured, and normalised to the total counts in the liver. The relative difference in the contribution to total liver function at each post-treatment time point to the pre-treatment time point was computed.

3. Results

Four of a planned 20 patients have consented to participate and have undergone pre-SABR SPECT imaging, ICG and elastography examinations. Two patients have completed 1 month and one patient has completed 6 month post-treatment imaging. Figure 1 shows the pre, and post treatment SPECT images for patients 1 and 3. Visual inspection of the pre and 1 month post-treatment SPECT images show decreased uptake in irradiated volumes at the 1 month time point. The 6 month time point for patient #1 shows slightly increased liver function compared with the 1 month time point in the low dose areas.

The contribution of normal liver receiving 5-50 Gy to the total liver function was analysed for patient #1. Figure 2 shows a decrease in contribution to liver function with increasing dose, the magnitude of which increased at the 6 month time point. A maximum of 41% reduction in liver function was observed at 6 months in the liver receiving 40-50 Gy. A potential compensatory increase in liver function in liver receiving less than 5 Gy was observed. It should be noted the total contribution of these liver volumes to the total liver function in the high dose regions is small; the 40-50 Gy volume contributed only 1.5% of the total liver function pre-treatment. This is a consequence of the SABR treatment technique, which has a very sharp dose fall off outside of the tumour which reduces dose to surrounding non-tumour liver.

IOP Conf. Series: Journal of Physics: Conf. Series **1154** (2019) 012009 doi:10.1088/1742-6596/1154/1/012009

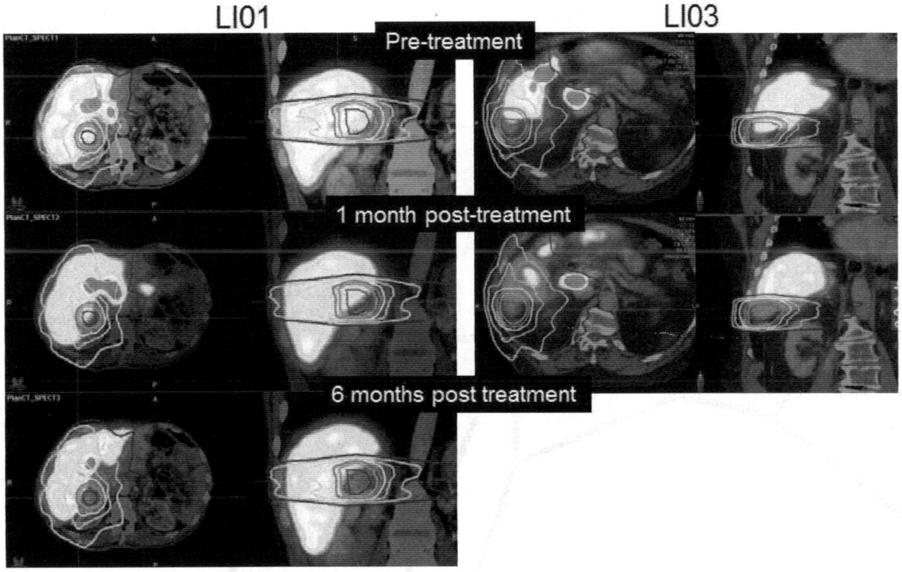

Figure 1. Pre-treatment, 1 month and 6 months post treatment 99m-Tc Mebrofenin SPECT scans for patients #1 and #3. Blue contours are the gross tumour volume. From orange to blue, the lines represent the 50, 40, 30, 20, 10 & 5 Gy isodose lines. The colourmap of the SPECT images have been normalised based on total liver counts.

4. Discussion

The current trial aims to improve understanding of liver function prior to and post-SABR by using a nuclear medicine tracer that is almost exclusively taken up by hepatocytes. There is a potential role for Tc-99m Mebrofenin SPECT in liver SABR pre-treatment in both primary liver cancer patients and patients with liver metastases from other cancers. Patients with primary liver cancer typically have heterogeneous liver function. Hepatocellular carcinoma however exhibits no dose response, that is, a high tumour dose is not necessarily required for tumour control [10]. The primary aim is thus to deliver sufficient dose to control the tumour whilst preserving liver function. Functional imaging may provide means to spare the highest quality regions of the liver. Patients with liver metastases will often have good liver function but are often undergoing multiple interventions prior to SABR such as surgery, transarterial-chemo-embolisation or radiofrequency ablation, which can reduce the volume of functional liver. Moreover liver metastases require a high biologically equivalent dose to achieve tumour control [10], and may present with multiple liver metastases. In this case pre-treatment functional imaging may provide a method to ensure sufficient future remnant liver after treatment.

The main limitation in the current workflow is registration uncertainties. The planning CT is an average of a 4DCT acquired on a flat couch top, whereas the CT component of the SPECT-CT is a snapshot on a curved couch-top. Rigid registration is a very good first approximation, given the small volumes involved in SABR planning, but residual uncertainties will still remain due to patient posture and inclusion of some blur due to breathing in the 4DCT. The SPECT image data however is acquired over a longer period of time (10 minutes), thus has respiratory motion blur contained in the image which should match the 4DCT average.

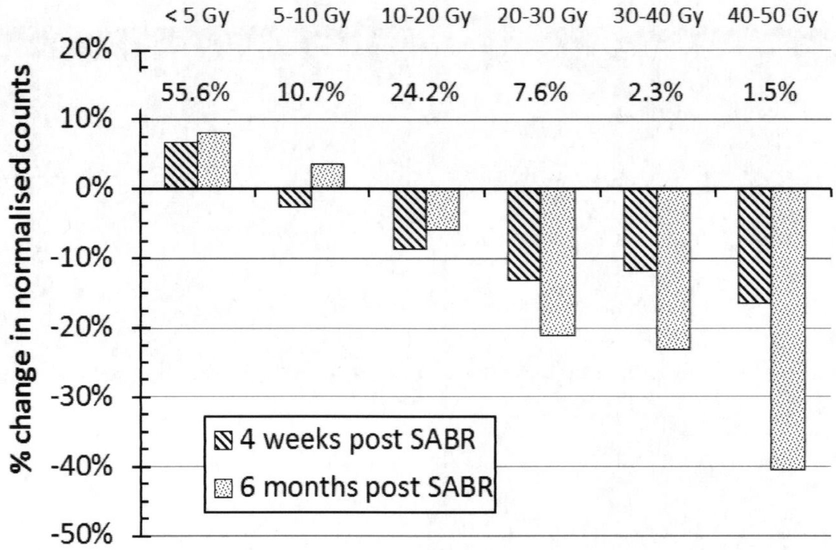

Figure 2. The change in contribution to the total liver function of the non-tumour liver receiving 5-50 Gy for patient #1. The percentages above each column pair are the pre-treatment contribution of each volume to the total liver function.

5. Conclusion

The Liver INSPECTR trial incorporates novel functional imaging to improve understanding of radiation effects on functional liver in the context of liver SABR. Initial results show reduction of liver function in regions of liver that receive high doses. Future work will further measure patient-specific changes in liver function post-SABR, and determine whether pre-treatment functional imaging can be incorporated into treatment planning to reduce dose to high functioning liver.

6. References

[1] Aitken K L and Hawkins M A 2015 *Clin Oncol* **27** 307-15
[2] Bujold A *et al* 2013 *J. Clin. Oncol.* **31** 1631-9
[3] Pan C C *et al* 2010 *Int. J. Radiat. Oncol. Biol. Phys.* **76** 94-100
[4] Bennink R J *et al* 2004 *J Nucl Med* **45** 965-71
[5] Bennink R J *et al* 2012 *Semin. Nucl. Med.* **42** 124-37
[6] Willowson K *et al* 2008 *Phys. Med. Biol.* **53** 3099-112
[7] Willowson K *et al* 2010 *Clin. Phys. Funct. Imag.* **30** 214-9
[8] Willowson K *et al* 2011 *Phys. Med. Biol.* **56** N145-52
[9] Bowen S R *et al* 2015 *Radiother. Oncol.* **115** 203-10
[10] Ohri N *et al* 2018 *Int. J. Radiat. Oncol.* DOI: 10.1016/j.ijrobp.2017.12.288

IOP Conf. Series: Journal of Physics: Conf. Series **1154** (2019) 012010 doi:10.1088/1742-6596/1154/1/012010

Use of contemporary prostate brachytherapy approaches in clinical trials

A Haworth[1], Y Sun[2], M Ebert[3,4], H Reynolds[2,5], J Betts[6], D Wraith[7], C Mitchell[5], D Murphy[2,5], B Parameswaran[5] and S Williams[2,5]
[1]School of Physics, University of Sydney, NSW Annette.haworth@sydney.edu.au
[2]Sir Peter MacCallum Dept. of Oncology, University of Melbourne, Melbourne, Vic
[3]School of Physics, University of Western Australia, WA
[4]Department of Radiation Oncology, Sir Charles Gairdner Hospital, WA
[5]Peter MacCallum Cancer Centre, Melbourne, Vic
[6]Faculty of Information Technology, Monash University, Vic
[7]Institute of Health and Biomedical Innovation, Queensland University of Technology, Brisbane, Qld

Email: Annette.haworth@sydney.edu.au

Abstract. A number of clinical trials are investigating the role of prostate focal brachytherapy as an alternative treatment to whole gland therapy, offering the opportunity for tumour dose escalation and/or reduced toxicity. Brachytherapy, either low dose rate or high dose rate, offers a treatment choice with both precision in dose delivery and opportunity for a highly conformal, non-uniform dose distribution. Whilst multiple consensus documents have published clinical guidelines for patient selection, there are insufficient data to provide clear guidelines for treatment planning approaches along with many other technical issues when practicing focal brachytherapy. Without consensus guidelines there is the potential for a diversity of practices to develop, leading to challenges in interpreting outcome data from multiple centres. We provide a quantitative framework for prostate focal brachytherapy that incorporates a personalized, biological approach that can be used as the foundation for future prostate focal brachytherapy trials.

1. Introduction

Prostate cancer (CaP) is globally the most commonly diagnosed cancer in men [1]. Randomised trials have demonstrated that radiotherapy is an effective form of treatment for localised CaP and that escalating the dose improves biochemical control [2-5]. Dose escalation however, also results in increasing doses of radiation being delivered to surrounding healthy tissue and is therefore limited by a threshold when rates of gastrointestinal and genitourinary toxicity become unacceptable [2]. Brachytherapy [BT] offers an opportunity to accurately deliver higher doses of radiation than can typically be delivered with external beam radiotherapy (EBRT) [6]. This is due to the high dose gradients that surround the radioactive source creating an opportunity to tightly conform the radiation dose to the target volume whilst controlling radioactive source placement [7]. Furthermore, when BT is delivered in a small number of high-dose fractions, the radiobiological benefits due to the low alpha/beta ratio for prostate cancer can be exploited [8]. BT dose escalation however, using a whole-gland approach is limited by normal tissue toxicity, in particular, urinary toxicity associated with dose to the urethra [9].

IOP Conf. Series: Journal of Physics: Conf. Series **1154** (2019) 012010 doi:10.1088/1742-6596/1154/1/012010

A number of studies have investigated "focal" or "focussed" approaches to BT treatments, whereby the aim is to deliver very high doses of radiation to only tumour bearing regions of the prostate, with lower doses delivered to the remainder of the gland in an effort to maximise tumour control and minimise side effects [10,11]. Whilst a number of clinical trials report promising results, heterogeneity in treatment approaches may lead to difficulties in interpreting inconsistencies or unexpected results. Furthermore, technical challenges may prohibit rapid translation to routine clinical care [12].

We propose a focal BT approach that applies quantitative imaging (radiomics) to not only standardise target volume definitions but also non-invasively quantify and classify tumour phenotype [13,14]. The biofocussed BT approach uses advanced imaging techniques to characterise the biological nature of tumours using a 3D, voxel-by-voxel spatial representation, described by Haworth et al [15]. Using a mathematical model, the radiation dose needed to destroy the tumour cells within each voxel can be calculated; with knowledge of the ideal dose distribution a treatment plan can then be generated using either a low-dose-rate (LDR) or high-dose-rate (HDR) approach. We hereby describe a framework for Biologically Optimised RadioTherapy (BiRT).

2. Methods

The framework for BiRT, that incorporates multiparametric MRI (mpMRI), is shown in Figure 1. The framework presents a personalised medicine approach to treatment whereby knowledge of the spatial distribution of tumour heterogeneity is first quantified following imaging. Treatment is then prescribed based on 3D maps of the spatial heterogeneity and a dose-painting approach [16]. The treatment is delivered under the control of image guidance techniques, and quantitative imaging at regular intervals following treatment is compared with the pre-treatment imaging data to monitor treatment response.

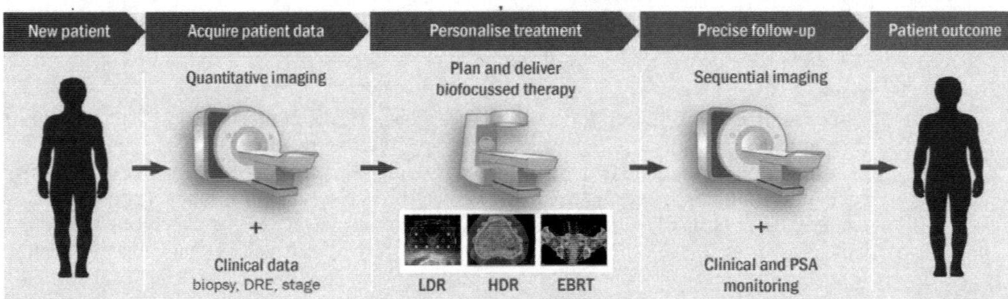

Figure 1. The BiRT framework

2.1 Acquire patient data

Multiparametric MRI is widely used in the staging and grading of newly diagnosed prostate cancer, and the PI-RADS system provides recommendations for scanning sequences and for reporting clinically significant cancers [17]. Regarding focal therapy, mpMRI may be used to guide biopsy, and has the potential to improve the detection of clinically significant cancers [18]. Similarly, mpMRI is the most commonly suggested method for defining the focal treatment volume within the prostate due to its high sensitivity and specificity [19]. Previous focal BT approaches however, have used an empirical or subjective approach to define target volumes, with definitions typically ranging from a geometric sub-division of the prostate (e.g. hemi-gland and sector-based approaches [20]), regions of positive biopsies [21] or a subjective delineation on MRI [22]. In contrast we suggest the use of computer aided detection (CAD) systems to delineate PCa from mpMRI [23,24] to improve consistency compared to manual delineation, which can suffer from inter-observer variability and low detection rates of satellite lesions [25]. A recent study by Sun et al for example, described the development of Gaussian kernel support vector machines (SVMs) and reported a prediction accuracy ranging from 70.4% to 87.1% as well as area under the curve (AUC) of the receiver operating characteristic (ROC) curve ranging from 0.81 to 0.94 [26].

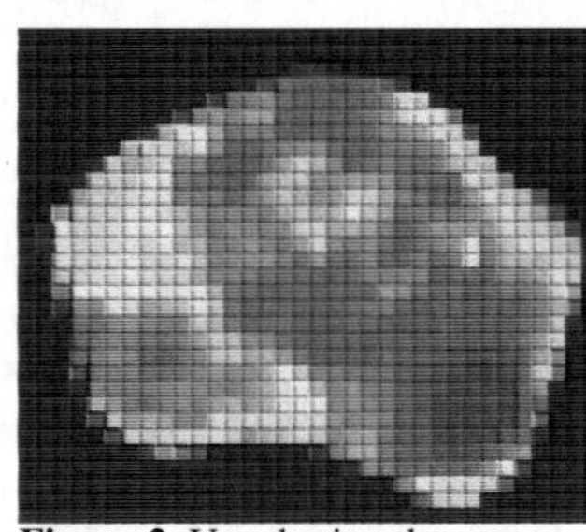

Figure 2. Voxel-wise characterisation of tumour biology to be used with biologically based inverse planning treatment methods [8].

IOP Conf. Series: Journal of Physics: Conf. Series **1154** (2019) 012010 doi:10.1088/1742-6596/1154/1/012010

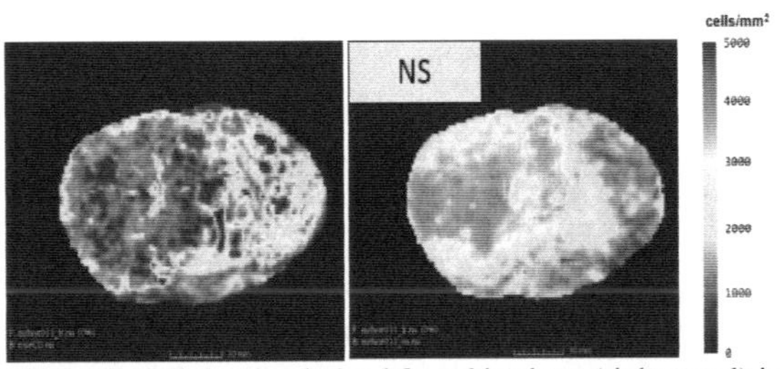

Figure 3. Cell density derived from histology (right panel) is compared with the predicted cell density derived using natural spline algorithm (left panel).

In addition to using mpMRI to define volumes for targeted treatments, mpMRI can also provide voxel-wise characterisation of the tumour phenotype [27]. Whilst radiogenomics analysis has been widely reported for its ability to capture intratumour heterogeneity in head and neck and lung cancers, the application in prostate cancer is less well developed. For the proposed biological approach to treatment planning, we have investigated the use of machine learning techniques to quantitate the spatial distribution of tumour cell density, tumour aggressiveness and hypoxia. Promising results indicate cell density prediction with a root mean square error of 1. 06 x 10^3 cells / mm^2, and differentiation of low and high-grade tumour with an AUC of 0.91 (Figure 2, *submitted*). Data were derived from a clinical study where in vivo mpMRI data was co-registered with histology using a sophisticated co-registration framework [28]. Mapping the spatial distribution of hypoxia in prostate cancer is challenging due to the difficulty in providing ground truth data from histology. Radiogenomics approaches offer apromising solution to this problem [29].

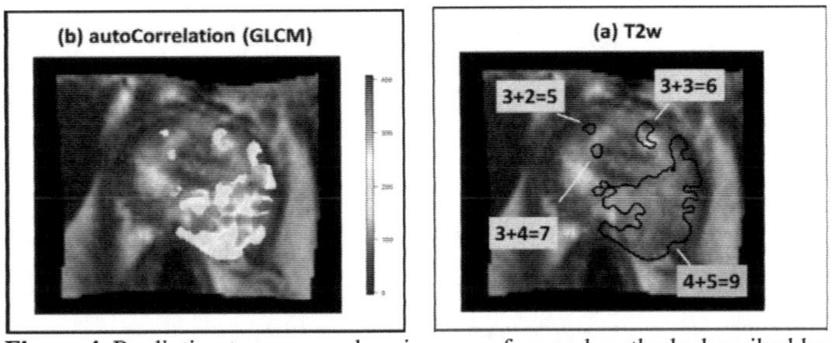

Figure 4. Predicting tumour grade using one of several methods described by Sun et al (submitted). GLCM: gray-level co-occurrence matrix. Left panel shows pathologist annotation overlaid on T2w MRI.

2.2 Personalised Treatment

With a voxel-wise representation of tumour location and characterisation of intratumour heterogeneity, treatment planning can be carried out using an inverse optimisation approach with biological objective functions. In the planning study of Haworth et al it was shown that by using an objective function based on tumour control probability (TCP), urethral doses could be reduced by up to eight-fold when compared with conventional, whole-gland treatment approaches and dose-based planning objectives [29]. It should be noted that in contrast to previous approaches to maximise radiation dose to the tumour bearing regions, the biological approach instead attempts to find the plateau on the TCP curve (Figure 3). With this approach, the normal tissue complication probabilities for each of the organs-at-risk can be minimised. Whilst we have shown that LDR treatment plans can be developed that are robust to seed placement uncertainty [30], more work in this area is required for HDR treatment planning. However, using the same principles, HDR treatments, when delivered using 3D source tracking technology to verify accurate delivery, offers the potential for a relatively simple but highly effective form of treatment for all prostate cancer risk groups.

2.3 Precise Follow-up

The current standard method for assessing treatment response following prostate radiotherapy is measurement of prostate specific antigen (PSA) [31]. However, its role in determining treatment response in BT is challenging due to the frequent "bounce" phenomenon [32], and its role for focused therapies is unclear. Instead, utilisation of mpMRI and PET imaging has the potential to offer improved outcome assessment after prostate therapy and has been previously reported by several researchers, though the approach is largely qualitative with conflicting results [33]. With the BiRT framework, we propose that quantitative biological and functional imaging methods will provide more reliable and reproducible methods for accurate assessment of PCa radiotherapy treatment response. Technical challenges include standardization of scanning protocols and quantitative measures to account for inter- and intra-scanner variability [34].

3. Conclusion

We propose a quantitative, biologically-based, image-guided approach to focal therapy using imaging protocols that could be standardly incorporated into clinical trials delivering outcomes that can be rapidly translated into clinical practice. This method provides an objective process, using imaging as a guide for mapping spatial tumour characteristics to the non-uniform dose distributions required for optimal treatment, realising the goal of voxel-level dose painting. The methods used to map those tumour characteristics from imaging data allow quantification and consideration of the underlying uncertainties. With this approach, BT offers an ideal treatment option, exploiting the ability to tightly conform dose distributions under controlled treatment delivery techniques.

4. Acknowledgements

This study was supported by NHMRC grant 1126955, PdCCRS grant 628592 with funding partners: Prostate Cancer Foundation of Australia, and the Radiation Oncology Section of the Australian Government of Health and Aging and Cancer Australia. Yu Sun is funded by the Melbourne International Research Scholarship, the Movember Young Investigator Grant through Prostate Cancer Foundation of Australia (PCFA) and Cancer Therapeutics Top-up Funding. Dr Reynolds is funded by the Movember Young Investigator Grant through (PCFA). The authors would like to thank Courtney Savill and Lauren Caspersz for their contribution in specimen preparation and MRI acquisition.

5. References

[1] Fitzmaurice C *et al* 2017 *JAMA Oncol.* **3** 524-48
[2] Denham J W *et al* 2015 *Radiother. Oncol.* **115** 301-7
[3] Eade T N *et al* 2007 *Int. J. Radiat. Oncol. Biol. Physics* **1** 682-9
[4] Zelefsky M J *et al* 2011 *Eur. Urol.* **60** 1133-9
[5] Dearnaley D P *et al* 2014 *Lancet Oncol.* **15** 464-73
[6] Hurley C *et al* 2006 *Nucl. Instrum. Meth. A* **565** 801-11
[7] Georg D *et al* 2014 *Int. J. Radiat. Oncol. Biol. Phys.* **88** 715-22
[8] Haworth A *et al* 2004 *Phys. Med. Biol.* **49** 3649-64
[9] Raleigh D R *et al* 2015 *Brachytherapy* **14** 795-800
[10] Valerio M *et al* 2017 *Eur. Urol.* **71** 17-34
[11] Langley S *et al* 2012 *BJU Int.* **109** Suppl 7-16
[12] Haworth A and Williams S 2017 *J. Contemp. Brachytherapy* **4** 383-9
[13] Jaffray D A *et al* 2015 *Semin. Radiat. Oncol.* **25** 292-304
[14] Kumar V *et al* 2013 *Magn. Reson. Imaging* **30** 1234-48
[15] Haworth A *et al* 2013 *Brachytherapy* **12** 628-36
[16] Ling C C *et al* 2000 *Int. J. Radiat. Oncol. Biol. Phys.* **47** 551-60
[17] Weinreb J C *et al* 2016 *Eur. Urol.* **69** 16-40
[18] Ahmed HU *et al* 2017 *Lancet* **6736** 1-8
[19] van der Heide U A *et al* 2012 *Magn. Reson. Imaging* **30** 1216-23
[20] Mason J *et al* 2015 *Radiother. Oncol* **117** 521-4
[21] Cosset J M *et al* 2013 *Brachytherapy* **12** 331-7
[22] Bauman G *et al* 2013 *Radiother. Oncol.* **107** 274-81
[23] Lemaître G *et al* 2015 *Comput. Biol. Med.* **60** 8-31
[24] Wang S *et al* 2014 *Biomed. Res. Int.* **2014** 789561
[25] Steenbergen P *et al* 2015 *Radiother. Oncol.* **115** 186-90
[26] Sun Y *et al* 2017 *Australas. Phys. Eng. Sci. Med.* **40** 39-49
[27] Aerts H J W L *et al* 2014 *Nat. Commun.* **5** 4006
[28] Reynolds HM *et al* 2015 *Med Phys.* **42** 7078-89
[29] Lyng H and Malinen E 2017 *Clin. Transl. Imaging* **5** 373-88
[30] Haworth A *et al* 2016 *Phys. Med. Biol.* **61** 430-44
[31] Betts J M *et al* 2017 *Procedia. Comput. Sci.* **108** 1522-31
[32] Roach M *et al* 2006 *Int. J. Radiat. Oncol. Biol. Phys.* **65** 965-74
[33] Paoluzzi M *et al* 2017 *Brachytherapy* **16** 1000-6
[34] Barchetti F and Panebianco V 2014 *Biomed. Res. Int.* **2014** 316272
[35] Jafar M M 2016 *World J. Radiol.* **8** 21

IOP Conf. Series: Journal of Physics: Conf. Series **1154** (2019) 012011 doi:10.1088/1742-6596/1154/1/012011

On the generation of synthetic CT for a MRI-only radiation therapy workflow for the abdomen

S-H Hsu[1], Q Peng[2] and W A Tomé[1]
[1]Department of Radiation Oncology, Montefiore Medical Center and Albert Einstein College of Medicine, Bronx, NY 10461, USA
[2]Department of Radiology, Montefiore Medical Center and Albert Einstein College of Medicine, Bronx, NY 10461, USA

E-mail: shhsu@montefiore.org

Abstract. MRI is an important imaging modality for contouring in radiotherapy due to its superior soft tissue contrast over CT, but it is not often used alone in a radiation therapy workflow because of the lack of electron density information which is required for dose calculations. This study investigates a method to generate synthetic CT from MRI to demonstrate the potential of using MRI alone in the radiation therapy workflow for abdominal targets. This method consists of (1) acquiring multiple MRI volumes, (2) applying intensity nonuniformity corrections to MR images, (3) identifying abdominal walls to distinguish air from bone and vertebral bodies to improve classifications between bone marrow and muscle tissues, (4) classifying tissues in each section using fuzzy c-means clustering with a spatial constraint and (5) assigning attenuation properties to individual tissue classes to generate synthetic CT. The quality of the synthetic CT was comparable to clinical CT. The proposed method allows automatic segmentation of tissue and generation of synthetic CT for the abdomen. Our results demonstrate the potential of using MRI alone to support treatment planning and image guidance for abdominal radiation therapy.

1. Introduction

MRI is increasingly being used in the radiation therapy workflow due to its superior soft tissue visualization, but it is not often used as a sole imaging modality in the planning process since the lack of electron density information precludes heterogeneity corrections. The use of multi-imaging modalities requires an imaging registration step and hence introduces uncertainties which are then propagated throughout the entire course of radiation therapy. To reduce these uncertainties, it has become of interest to streamline the workflow by implementing a planning process that is solely based on MRI. Several methods have been investigated in order to generate electron density maps from MRI for treatment planning, such as methods that manually identify tissues types and assign bulk densities [1, 2], atlas-based methods to automate the process in generating electron density maps [3, 4], and voxel-based methods to account for individual differences that are typically lacking in atlas-based methods [5, 6]. These studies have shown promising results in terms of pseudo-CT quality, and demonstrated the potential of using MRI alone in the workflow for brain and prostate radiation therapy.

Due to breathing and gastrointestinal motion [7], the registration uncertainty is likely larger in the abdomen than other body sites. Moreover, superior tissue contrast is important to identify liver and pancreatic tumours, which means that the patients with abdominal lesions could benefit greatly from using MRI as the sole image modality throughout their treatment course. Therefore, this study investigates a method of generating synthetic CT to support the delineation of targets and organs-at-risk (OARs) for the use in treatment planning, dose calculations, and the use of synthetic CT or MRI for treatment verification towards a MRI-only radiation therapy workflow.

2. Methods and materials

To explore the feasibility of synthetizing CT from MRI to support treatment planning, diagnostic MRI scans with gadolinium (Gd) contrast were analysed. These image volumes included in-phase, fat and water images resulting from a mDixon sequence and T2-weighted images using a T2-turbo spin echo sequence (cf. Figure 1). These images were corrected for their nonuniform intensity using the N4itK algorithm implemented in SLICER (Version 4.6, surgical processing laboratory, Brigham and Women's Hospital, Boston, MA).

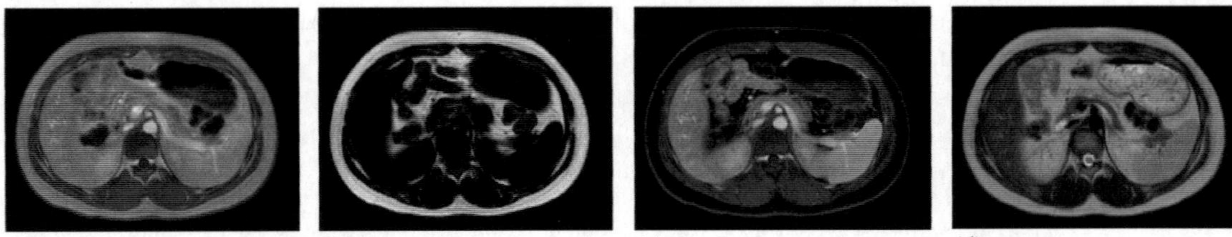

(a) (b) (c) (d)

Figure 1. (a) in-phase, (b) fat, (c) water and (d) T2-weighted MR images.

Due to ultra-short T2 and T2* of cortical bone, bone cannot be easily distinguished from air on conventional MRI. To address this issue without using ultra-short echo time (UTE) pulse sequences, the abdominal wall was automatically identified using intensity thresholding and morphological processing on in-phase images, and two compartments were then created, one compartment containing air pockets and lungs and the other compartment containing bone (i.e. ribs, sternum and spine). In the outer compartment (that included bone), the MR intensity of muscle tissues were close to bone marrow on our studied MR images. Considering CT number differences between muscle and bone marrow, it may be necessary to distinguish these two tissue types. Therefore, the vertebral bodies were automatically identified based on their known spatial location in the abdominal MR images using cylindrical shapes. Hence, in this pre-processing step, four sections were created: (1) the air section from the inner abdomen compartment only containing air pockets, (2) the solid bone section from the outer abdomen compartment containing bone only, (3) the vertebral body section from the outer abdomen compartment and (4) the residual tissue section including all remaining tissues in both compartments.

A Hounsfield Unit (HU) value of -1000 was assigned to the air section with a tissue probability of 1 while a HU value of 1000 was assigned to the solid bone section with a tissue probability of 1. Using in-phase and water images a probabilistic tissue classification using fuzzy c-means clustering (FCM) with a spatial constraint as described in Hsu et al [5] was applied to the two remaining sections. The FCM algorithm assigns each voxel in an image a certain probability of belonging to a given tissue class, i.e. each voxel has a certain probability of being a member of each of the tissue classes. In the residual tissue section, four tissue probability maps were generated, which represented the classes of fat, vessels (and Gd contrast regions), muscle and internal organs. In the vertebral body section, three tissue probability maps were generated, representing the classes of bone marrow, inter-vertebral space and vessels. Corresponding CT numbers were then assigned to individual probability maps with HUs of -100 for fat, 50 for vessels, 70 for muscle and internal organs, 100 for the inter-vertebral space and 250 for bone marrow. Then, the probability-weighted sum of the attenuation properties of each voxel yields the synthetic CT.

The synthetic CT was then imported into a commercial treatment planning system (Eclipse 11.0, Varian, Palo Alto, CA). To mimic treatment planning in the abdomen, targets and OARs were contoured on in-phase images. The pancreas was contoured as clinical target volume (CTV) and the planning target volume (PTV) was generated using an isotropic 1 cm expansion with an additional expansion 0.5 cm superior/inferior directions. Planning was performed on the synthetic CT employing 2-arc volumetric modulated arc therapy (VMAT) with a total dose of 50.4 Gy in 28 fractions. Dose calculations were performed using the analytical anisotropic algorithm (AAA) with and without density corrections on synthetic CT.

3. Results

Figure 2 shows four probability images generated using FCM classification. The resulting synthetic CT is shown in Figure 3. Some misclassifications were found at tissue interface and regions with nonuniform intensities, but the quality of the synthetic CT was comparable to clinical CT. Comparing dose volume histograms (DVHs) with and without density corrections, no significant difference was found for OARs, but the dose to the 95% and 99.9% volume (D95% and D99.9%) for PTV was ~1% lower for the plan without density corrections.

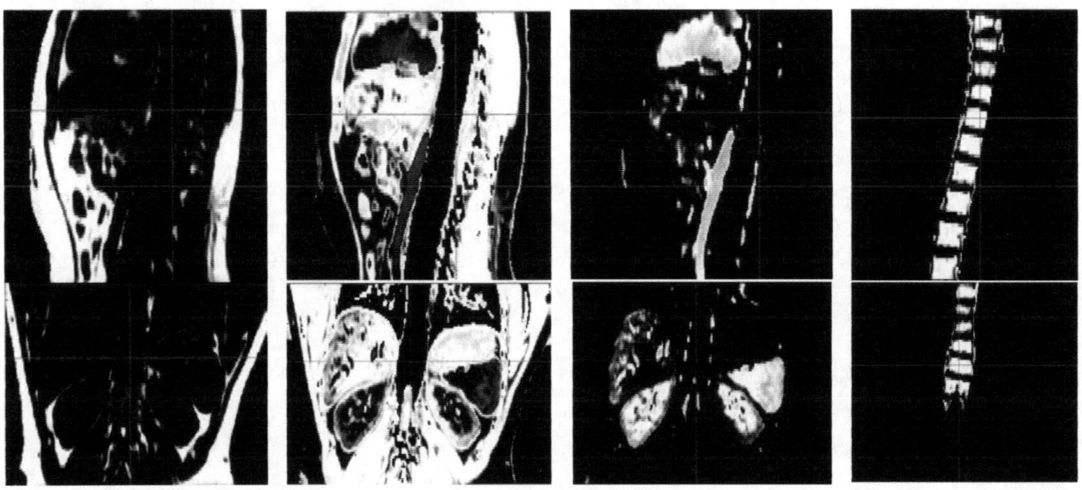

(a) (b) (c) (d)

Figure 2. Four tissue classes in sagittal (upper row) and coronal views (lower row): (a) fat, (b) muscle and internal organs, (c) vessels and contrast regions and (d) bone marrow.

4. Discussion

Our proposed method allows automatic segmentation of tissue and the generation of synthetic CT for the abdomen. In particular our method reduces the possibility of mislabelling air, bone and/or fluid and therefore yields improved CT number accuracy for bone marrow and spinal cord using conventional MRI sequences. However, there were residual nonuniform intensities within given tissue types on MR images, which resulted in tissue misclassification of voxels within those tissue types when generating the synthetic CT. The diagnostic MRI sequences used in this study have not been optimized for treatment planning purposes, so the quality of the generated synthetic CT is still limited. In addition, the CT numbers assigned to individual tissue classes were estimated based on the values of the CT dataset for one clinical case. Investigating what the adequate CT numbers for each tissue type are will be necessary to further improve the quality of synthetic CT.

Dose volume metrics only differed ~1% in target coverage between the plans with and without density corrections for the case studied. This small difference was likely due to the simulated target being far from the lungs and the use of an arc delivery technique. When abdominal targets are close to the lungs, generating synthetic CT to allow for density corrections may be more important for improving dosimetric accuracy.

Our current results clearly demonstrate the potential of using MRI alone to support treatment planning and image guidance for treatment verification in abdomen. Future work will include

IOP Conf. Series: Journal of Physics: Conf. Series **1154** (2019) 012011 doi:10.1088/1742-6596/1154/1/012011

improving the quality of the synthetic CT by optimizing the MR imaging sequences with sufficient tissue contrast to separate required tissue types, evaluating the robustness of the proposed method for diverse cases and its dosimetric accuracy in comparison to regular CT, determining its limitations, and exploring the proper workflow in treatment verification using synthetic CT or MRI.

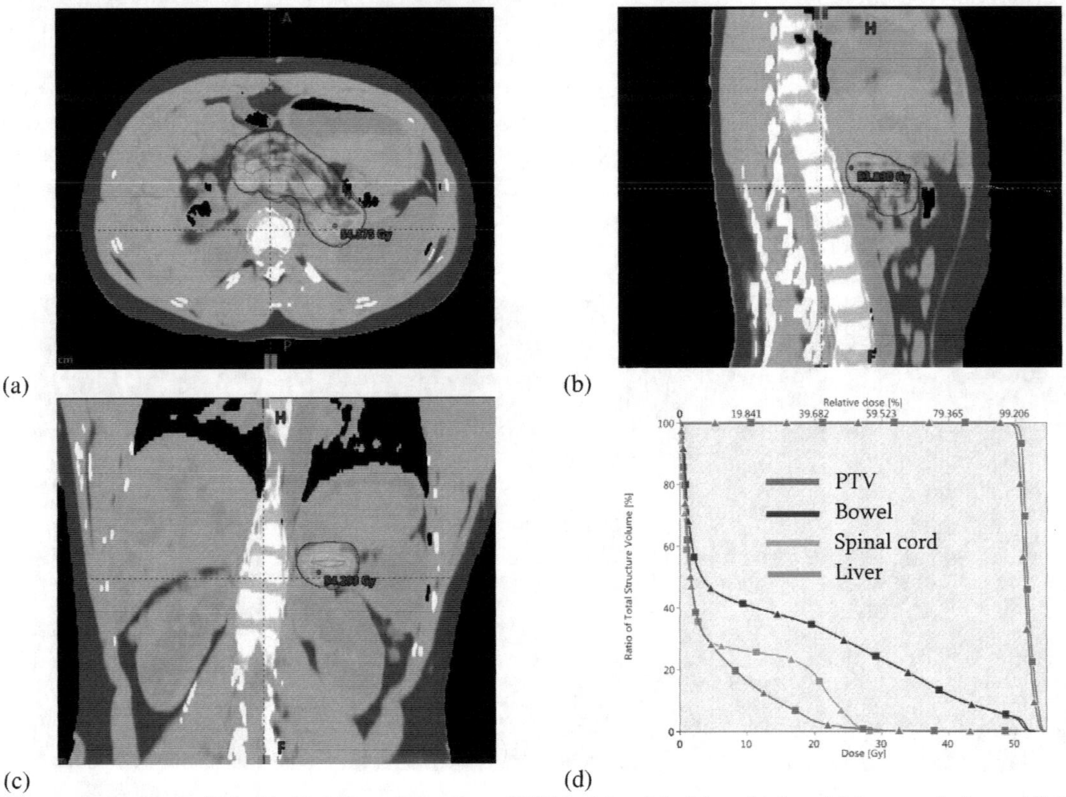

Figure 3. Synthetic CT with 50.4 Gy colorwash on PTV in (a) axial, (b) sagittal and (c) coronal views. (d) DVHs for synthetic CT with density corrections (■) and without corrections (▲).

5. Acknowledgements

This work was supported by a research grant from Varian Medical Systems, Inc.

6. References

[1] Jonsson J H *et al* 2010 *Radiat. Oncol.* **5** 62
[2] Lee Y K *et al* 2003 *Radiother. Oncol.* **66** 203-16
[3] Dowling J A *et al* 2012 *Int. J. Radiat. Oncol. Biol. Phys.* **83** e5-11
[4] Sjolund J *et al* 2015 *Phys. Med. Biol.* **60** 825-39
[5] Hsu S H *et al* 2013 *Phys. Med. Biol.* **58** 8419-35
[6] Johansson A *et al* 2011 *Med. Phys.* **38** 2708-14
[7] Kyme A Z *et al* 2011 *PLoS One* **6** e21727

IOP Conf. Series: Journal of Physics: Conf. Series **1154** (2019) 012012 doi:10.1088/1742-6596/1154/1/012012

3D sensitive volume microdosimeter with improved tissue equivalency: charge collection study and its application in ^{12}C ion therapy

B James[1], L T Tran[1], D Bolst[1], D Prokopovich[2], M Reinhard [1,2] M Lerch[1], M Petasecca[1], S Guatelli[1], M Povoli[3], A Kok[3], N Matsufuji[4], M Jackson[1,5] and A Rosenfeld[1]
[1]Centre for Medical Radiation Physics, University of Wollongong
[2]NSTLI Nuclear Stewardship, Australian Nuclear Science and Technology Organization, Australia
[3]SINTEF, Norway
[4]National Institutes for Quantum and Radiological Science and Technology, Chiba, Japan
[5]University of New South Wales, Sydney NSW 2052, Australia

E-Mail: anatoly@uow.edu.au

Abstract. This research focuses on the characterisation of a new 3D sensitive volume (SV) microdosimeter covered with polyimide – a material which closely mimics human tissue. The electrical and charge collection properties of the device were investigated and its application in ^{12}C ion therapy were studied. Charge collection studies revealed uniform charge collection and no cross talk between adjacent SVs. To study the microdosimetric response in ^{12}C ion therapy, the new polyimide mushroom microdosimeter were placed at various positions along the central axis of a 290 MeV/u ^{12}C ion spread out Bragg peak (SOBP) at the Heavy Ion Medical Accelerator in Chiba (HIMAC), Japan. From these microdosimetric spectra, dose mean lineal energy ($\overline{y_D}$) and RBE_{10} results were obtained, with RBE_{10} increasing from 1.3 at the entrance to 2.7 at the end of the SOBP. The results obtained in this work show that the new generation of mushroom microdosimeters, covered with tissue equivalent polyimide material, are a useful tool for quality assurance in heavy ion therapy applications.

1. Introduction

Particle therapy has many advantages over conventional photon therapy, particularly for treating deep-seated solid tumours due to its greater conformal energy deposition achieved in the form of the Bragg Peak (BP). Successful treatment with heavy ions requires knowledge of the relative biological effectiveness (RBE) of the primary and secondary radiation field. Different methods and approaches are used to calculate the RBE-weighted absorbed dose in heavy ion therapy. The RBE calculated using the microdosimetric approach been reported [1] for a therapeutic ^{12}C beam using a tissue equivalent proportional counter (TEPC). However, the large size of commercial TEPCs can cause the RBE values to be smeared dramatically close to and in the distal part of the BP, which may have significant clinical impact. The Centre for Medical Radiation Physics (CMRP), University of Wollongong, has initiated the concept of silicon microdosimetry to address the shortcomings of the TEPC [2]. In the course of

IOP Conf. Series: Journal of Physics: Conf. Series **1154** (2019) 012012 doi:10.1088/1742-6596/1154/1/012012

this research, a new 3D SV microdosimeter covered with a tissue equivalent material in the process of fabrication has been investigated and its application in ^{12}C therapy has been studied.

2. Method

2.1 Detector Structures

The Mushroom microdosimeter structure used in this work is called trenched planar, this consists of 3D cylindrical SVs with a planar n+ core produced by ion implantation (planar technology). Each SV is surrounded with a complete p+ doped trench filled with polysilicon (Fig 1b). SEM images of this device's structure show that a complete trench was etched (Fig. 1a) and the trench was filled with polysilicon (showed in Fig. 1b).

The structure is based on an array of 3D cylindrical SVs with diameters of 18 μm or 30 μm, fabricated on a high resistivity p-SOI with a 10 μm thick active layer bonded to a low resistivity supporting wafer and 2 μm silicon oxides in between. The SVs are separated into odd and even arrays. A P-stop layer has been deposited everywhere on both device structures and under the pad to avoid metal-oxide-semiconductor (MOS) build up charge effect under the metal buses [3].

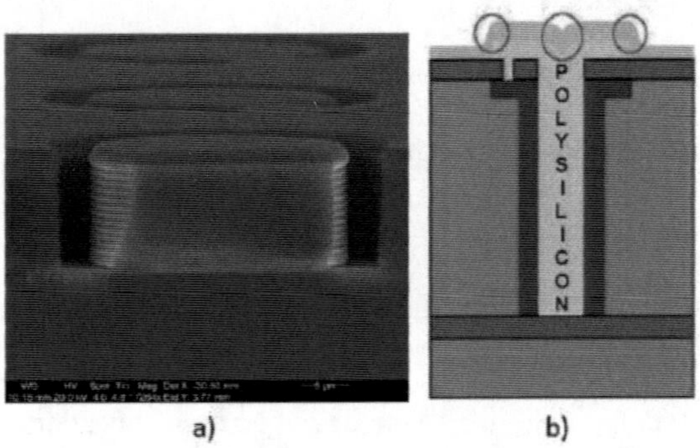

Figure 1. a) SEM image of a trench planar mushroom microdosimeter with a complete trench, b) trench was doped with p+ and filled with polysilicon [3].

The detector utilises the fact that even and odd rows of SVs are read out independently to avoid events in adjacent sensitive volumes being read as a single event in the case of oblique charged particle tracks. The charge collection and response of these detectors has been thoroughly investigated and documented in previous works [3], [4]. The 3D Mushroom microdosimeter was connected to a low noise spectroscopy-based readout circuit, allowing lineal energy measurements as low as 0.15 keV/μm in tissue [3].

In order to improve the tissue equivalence of the new microdosimeters, they have been covered with a 12 μm layer of polyimide, as seen in Figure 2. While the polyimide layer will serve to improve the tissue equivalence of the microdosimeter, charge collection and uniformity studies were required to understand how the microdosimeter output would be affected by this new layer.

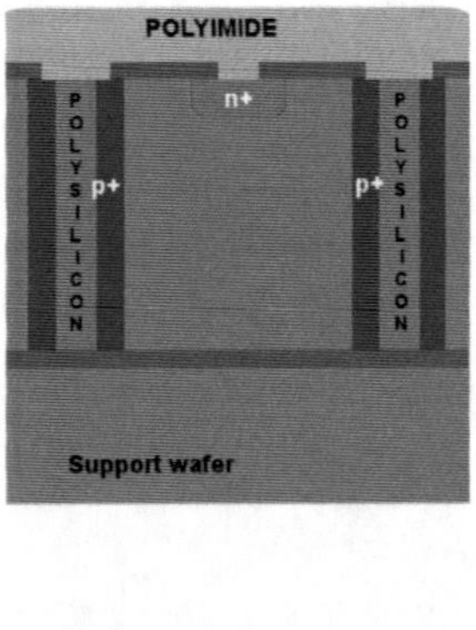

Figure 2. Trenched planar mushroom microdosimeter, where the trench is filled with polysilicon p+ column, then overlayed with the 12 μm polyimide layer.

IOP Conf. Series: Journal of Physics: Conf. Series **1154** (2019) 012012 doi:10.1088/1742-6596/1154/1/012012

2.2 Charge Collection Study

The Mushroom microdosimeters was studied using the ion beam induced charge collection technique (IBIC) at the 6 MV accelerator SIRIUS, located at the Centre for Accelerator Science (CAS) facility at ANSTO. This system includes a Confocal Heavy Ion Micro-Probe (CHIMP) which is capable of delivering different types of ions such as Carbon, Helium and Hydrogen ions. This beamline is widely used for analysing and characterising samples using ion beam analysis (IBA) and produces high-current and high brightness ion beams with exceptional energy resolution [5].

Figure 3. 6 MV accelerator SIRIUS, located at the Centre for Accelerator Science (CAS) facility at ANSTO [5].

The IBIC measurements utilized a microbeam of 5.5 MeV He^{2+} ions which were raster scanned over the surface of the sample. Energy deposited in the microdosimeter was measured using an AMPTEK A250 charge sensitive preamplifier and a Canberra 2025 Shaping Amplifier with 1μs shaping time. The signal corresponding to the beam's position as well as the charge collection for each event was processed into an event-by-event list mode file. The data was processed into median charge collection image maps for spatial correlation of the energy deposition of the scanned area [3]. The energy calibration was performed using a calibrated pulse generator which was calibrated with a 300μm thick planar silicon fully depleted PIN diode with 100% Charge Collection Efficiency (CCE) in response to 5.5 MeV He^{2+} ions.

2.3 HIMAC Experiment

To study its performance in ^{12}C therapy, the 3D mushroom microdosimeters were irradiated at the Heavy Ion Medical Accelerator in Chiba (HIMAC), located at the National Institute of Radiological Science (NIRS) in Japan. The facility consists of ion sources, a radio frequency quadrupole linear accelerator (LINAC) for low speed ions and an Alvarez LINAC for medium speed ions as an injector to the two synchrotrons, with maximum energy of 400 MeV/u [4].

Figure 4. Panorama showing the biological sciences beamline at the Heavy Ion Medical Accelerator in Chiba (HIMAC), Japan.

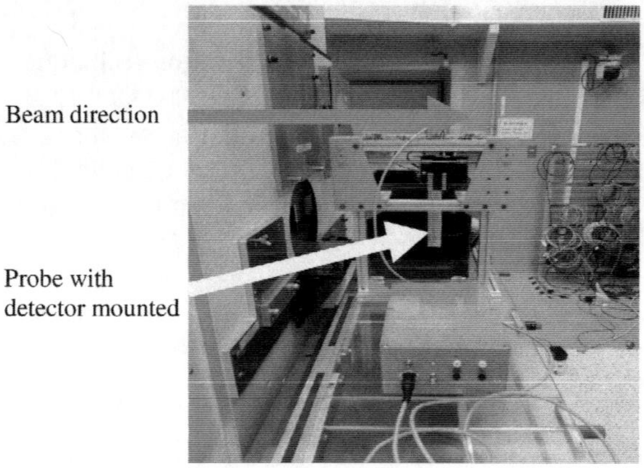

Figure 5. The remote control movable platform holding the detector mounted on the MicroPlus probe, connected to power supply box.

A 60 mm spread out Bragg peak (SOBP) produced with 290 MeV/u ^{12}C ions was used in this experiment. The polyimide mushroom microdosimeters were connected to a low noise spectroscopy-based readout circuit (also called a MicroPlus probe). The probe was then placed in a water tank, Figure 5, and measured at different depths in water along the central axis of the SOBP of the 290 MeV/u ^{12}C ion beam.

3. Results and Discussion

3.1 Charge Collection Study

Figure 6 shows the MCA spectra obtained at 0V and 10V for the Mushroom microdosimeter, using the IBIC technique. It can be seen that at 0V the detector is almost completely depleted, with only a 2% difference in CCE between 0 and 10V. Overall the trenched planar structure has been shown in previous work to have a CCE of 96% [3].

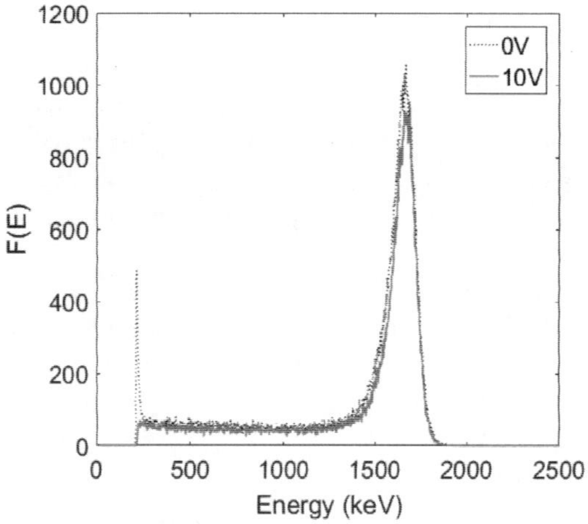

Figure 6. MCA spectra obtained during IBIC for 0V and 10V, for the same scan size of the Mushroom microdosimeter, only even array is connected.

Figure 7 shows the median energy map of multiple sensitive volumes connected in the even arrays. It can be seen that each of the sensitive volumes has very uniform charge collection and well-defined volume shapes. Very low energy charge was observed in narrow region surrounding each SV due to the trench wall doped with p+. Furthermore, no charge collection can be seen within the connecting regions between sensitive volumes, or in the unconnected (odd) array.

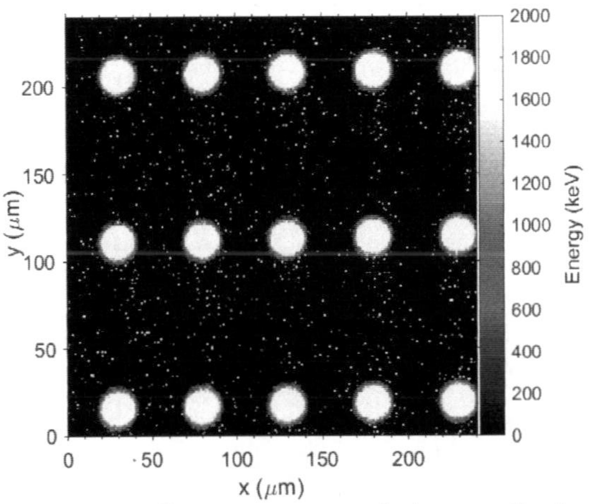

Figure 7. Median energy map of charge collection within three rows of the Mushroom microdosimeter at 10V, only even array is connected.

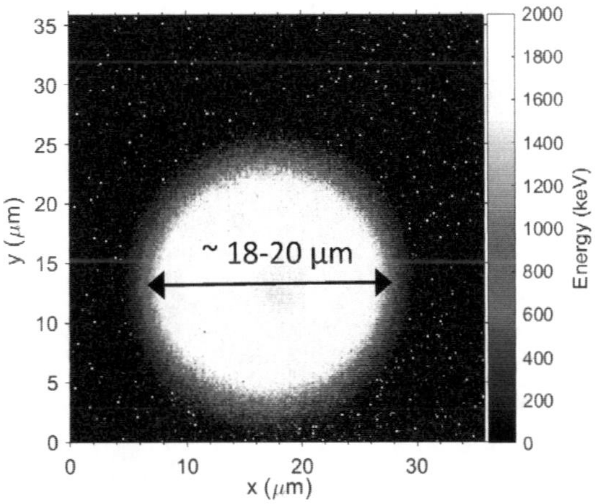

Figure 8. Median energy maps generated using a detector bias of 10V, showing a single SV.

Reducing the IBIC scan size, a single SV can be viewed in Figure 8. This reveals exceptionally uniform CCE within the SV, and a slight reduction of energy deposition was observed in the n+ core region of the SV (approximately 80% CCE compared to the SV). It also confirms that the size of a single SV is ~18-20μm.

3.2 Microdosimetric measurements of Carbon ions - HIMAC

Figure 9 shows microdosimetric spectra and $\overline{y_D}$ values obtained at different depths in water, along the central axis of the ^{12}C SOBP. At the entrance, $\overline{y_D}$ is approximately 15 keV/μm, as the depth in water increases the microdosimetric spectra obtained shifts to the right, and consequently the lineal energy deposition increases, as the primary beam energy decreases. This leads to higher energy deposition reaching a maximum at the distal part of the SOBP (~148 mm), where the $\overline{y_D}$ measured was 195 keV/μm.

The derived RBE_{10} values based on the microdosiemtric kinetic model (MKM), obtained in the SOBP are shown in Figure 10. The value of RBE at the entrance was calculated to be from 1.3 and increased to 2.7 at the end of the SOBP. The experimental data obtained with the microdosimeter can be seen to match the TEPC data extremely well in the entrance and at the end of the SOBP. Near the end of the SOBP (~148 mm), the high spatial resolution of the SOI mushroom microdosimeter enables more detailed RBE_{10} measurements to be obtained at the end of the SOBP than the TEPC. The higher experimental values obtained downstream of the SOBP can be attributed to the PMMA sheath, resulting in a different spectra of secondary particles than that observed by the TEPC which were carried out in water.

IOP Conf. Series: Journal of Physics: Conf. Series **1154** (2019) 012012 doi:10.1088/1742-6596/1154/1/012012

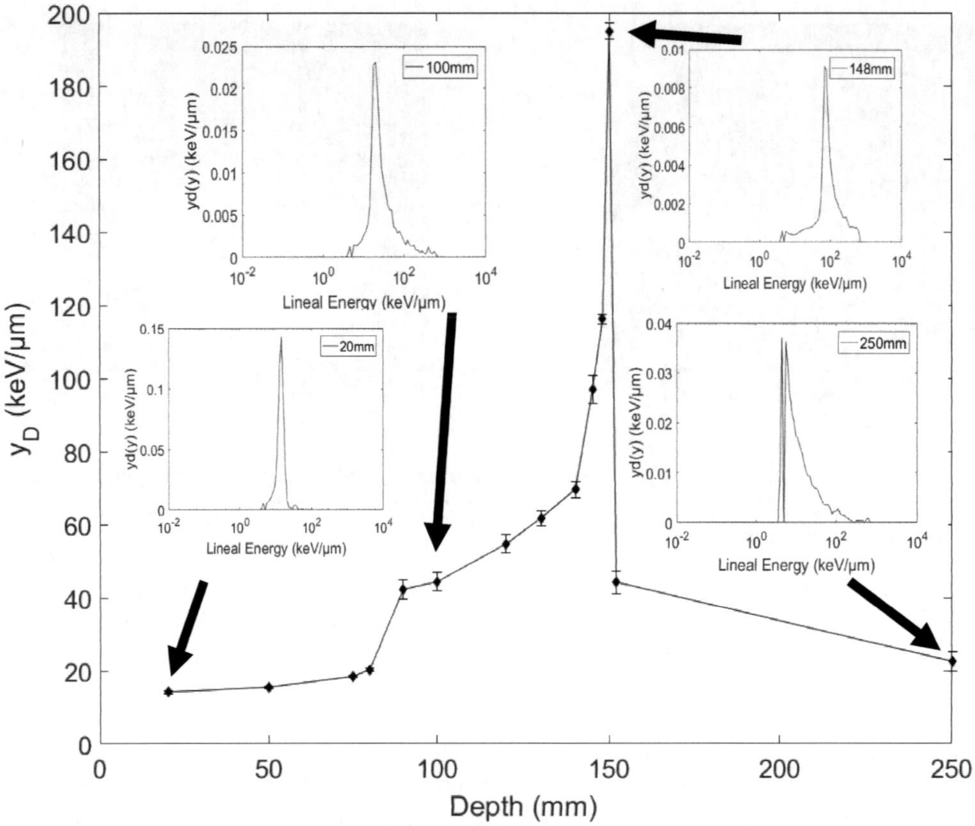

Figure 9. Microdosimetric spectra obtained by Polyimide mushroom microdosimeter in response to 290 MeV/u SOBP 12C ions at HIMAC.

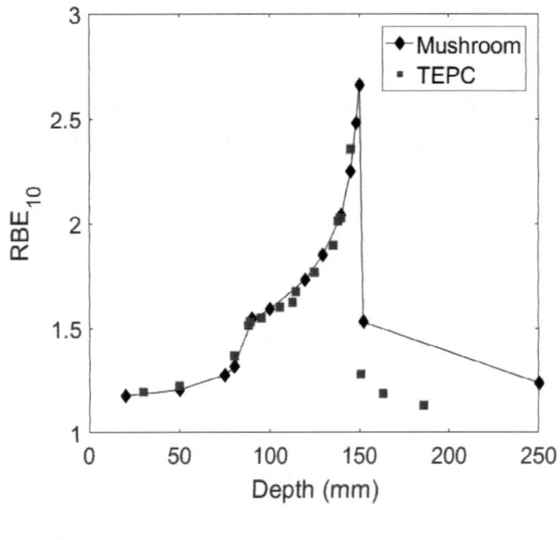

Figure 10. RBE_{10} distribution obtained by Polyimide mushroom microdosimeter in response to 290 MeV/u SOBP ^{12}C ions at HIMAC. Compared with TEPC data taken from Y. Kase [1].

4. Conclusion

The charge collection of a 3D microdosimeter with a polyimide covering has been studied using the IBIC technique and revealed a uniform charge collection with no cross-talk between SVs. A range of microdosimetric measurements were then able to be obtained using the new mushroom microdosimeter. Giving a maximum $\overline{y_D}$ and RBE_{10} of 195 keV/um and 2.7, respectively. We demonstrated that coating mushroom SVs with polyimide layer in the process of fabrication does not change electrical parameters and charge collections in SVs. Currently this version of mushroom is only covered with a 12 μm polyimide layer on top of the device, future versions of the mushroom microdosimeter will be fully etched silicon surrounding the sensitive volume and then filled with polyimide, further increasing the tissue equivalence of the device when considering the secondary particle production within the device. These results demonstrate the incremental improvement of the design and manufacturing of silicon microdosimeters with improved tissue equivalency.

5. References

[1] Kase Y *et al* 2011 *J. Radiat. Res.* **52** 789-96
[2] Tran L T 2015 *IEEE T. Nucl. Sci.* **62** 504-11
[3] Tran L T *et al* 2017 *IEEE T. Nucl. Sci.* **65** 467-72
[4] Tran L T *et al* 2015 *IEEE T. Nucl. Sci.* **62** 3027-33
[5] Pastuovic Z *et al* 2017 *Nucl. Instrum. Meth. B* **404** 1-8

IOP Conf. Series: Journal of Physics: Conf. Series **1154** (2019) 012013 doi:10.1088/1742-6596/1154/1/012013

Development of convergent close-coupling approach to hadron interactions with matter

A S Kadyrov, I B Abdurakhmanov, Sh U Alladustov, J J Bailey and I Bray
Curtin Institute for Computation and Department of Physics and Astronomy, Curtin University, GPO Box U1987, Perth 6845, Australia

E-mail: A.Kadyrov@curtin.edu.au

Abstract. Activities undertaken at Curtin University towards developing a comprehensive approach to all aspects of hadron interactions with matter are reviewed.

1. Introduction

One of the primary benefits of hadron therapy is a large ratio of the radiation dose at the Bragg peak to that in the plateau region before the peak. From the physics point of view three factors must be considered that may potentially enhance or undermine this benefit. (i) In order to reach deep-seated tumour cells therapeutic ion beams should have energies from a hundred up to a few hundreds of MeV per a.m.u. At such high energies, energy losses due to ionisation are small. The main channels of energy losses here are nuclear collisions, which may lead to fragmentation of nuclei of the target, and of the incident beam particles if the latter is a heavy ion like ^{12}C [1]. These also result in emission of neutrons. In the plateau region, the nuclear fragments and emitted neutrons can cause side effects, including new tumour cells. In addition, the nuclear fragmentation processes may disperse the beam and substantially affect the composition of the beam reaching the Bragg peak region. This complicates radiation dose simulations, which usually assume the initial beam composition. Due to complexity of nuclear interactions, there is no unified theory capable of describing them. Interaction models are generally anchored to experimental data, which in some areas are far from complete [2]. (ii) As the residual beam particles approach the Bragg peak region, their energy reduces significantly. When the energy falls below tens of MeV, energy losses due to excitation and ionisation of the target atoms and molecules become dominant. It is often stated that the theory of these processes is well developed. However, this is not quite correct. Currently available Monte Carlo simulation packages for modelling of biological damage induced by hadrons rely on the Bethe-Bloch (BB) theory of energy losses. However, the BB theory and its numerous improvements cannot meet accuracy requirements. There are a number of reasons for this. First, the BB-type theories do not include electron-capture channels. Second, they are perturbative and their applicability is limited to projectile energies higher than several MeV. Furthermore, they are based on the so-called dipole approximation to the first Born approximation. Third, the BB-type theories do not include coupling between various reaction channels. When hadrons approach the Bragg peak region, their energy diminishes considerably. In this energy region, the probability of electron capture becomes dominant and coupling between the channels – significant. In fact, when the energy falls below 100 keV per a.m.u., further energy losses are purely due to electron capture. Hence, accurate simulations of ion beam passage through matter at the micro- and nanoscale require inclusion of the electron-capture channels in addition to ionization [3]. (iii) In the Bragg peak region, the probability of target ionisation by the incident hadrons reaches its maximum. This leads to production of a copious number of secondary electrons. These so-called δ-rays have energies sufficient to further ionise the target and, as a result of multiple scattering, are capable of causing DNA double-strand breaks. It is estimated that among all the double strand breaks of DNA molecules of tissue treated by ion therapy, about 70% are produced by the δ-electrons [3]. Therefore, it is necessary to understand interactions of the produced δ-electrons in combination with interactions of the beam ions.

Thus, detailed knowledge of hadron beam interactions with atoms and molecules is paramount for accurate dose simulations in hadron therapy. We have been working on a comprehensive approach to modelling hadron interactions with matter that is capable of addressing all three factors mentioned above. We have developed an approach to hadron collisions with atomic and molecular targets. The approach known as the convergent close-coupling (CCC) [4-6] accounts for all possible reaction channels including ionisation and electron capture and is capable of providing the most detailed and accurate information about the collision process. We have also developed the CCC approach to electron-impact ionisation of atomic and molecular targets [7] applicable to δ-electrons. In addition, work is in progress on developing CCC-type approach to nuclear fragmentation, building on our experience in modelling nuclear reactions (see [8-11] and references therein). Further we concentrate on one aspect of these activities. We give a brief account of the CCC approach to hadron scattering in the Bragg peak region.

2. Development of convergent close-coupling approach to hadron-atom collisions

For simplicity we consider hadron collisions with a single-electron target. Generalisation for more complex targets is described in [12-14]. The total scattering wave function is a solution to the fully quantum-mechanical Schrödinger equation

$$(H - E)\Psi_i^+ = 0, \tag{1}$$

where H is the three-body Hamiltonian and E is the total energy. Conventional approaches to ion-atom collisions are based on the approximate time-dependent Schrödinger equation for the electronic part of the wave function that follows from Eq. (1) in a semiclassical approximation [15]. The CCC approach is different and based on expansion of the three-body wave function as

$$\Psi_i^+ \approx \sum_{\alpha=1}^{N} F_\alpha(t, \boldsymbol{b}) e^{i\boldsymbol{q}_\alpha \cdot \boldsymbol{\rho}_\alpha} \psi_\alpha(\boldsymbol{r}_\alpha) + \sum_{\beta=1}^{M} G_\beta(t, \boldsymbol{b}) e^{i\boldsymbol{q}_\beta \cdot \boldsymbol{\rho}_\beta} \psi_\beta(\boldsymbol{r}_\beta), \tag{2}$$

where $\boldsymbol{b}$ is the impact parameter. Index α denotes a quantum state in a channel where the projectile of relative momentum $\boldsymbol{q}_\alpha$ is incident on a bound state of the target atom. Index β is reserved for the rearrangement channel, where the atom formed by the projectile after electron capture has momentum $\boldsymbol{q}_\beta$ relative to the stripped target nucleus. The position of the projectile with respect to the centre of mass of the target is denoted by $\boldsymbol{\rho}_\alpha$, while $\boldsymbol{\rho}_\beta$ is the position of the projectile-electron pair with respect to the target nucleus. The position of the electron relative to the target proton is $\boldsymbol{r}_\alpha$, while $\boldsymbol{r}_\beta$ is the electron position relative to the projectile. The basis includes the continuum in a suitably discretized form, where ψ_α and ψ_β are the target and projectile-centred pseudostates, N and M are the numbers of basis functions on the target and projectile centers. The time-dependent expansion coefficients $F_\alpha(t, \boldsymbol{b})$ and $G_\beta(t, \boldsymbol{b})$ at $t \to \infty$ represent the transition amplitudes into the target and projectile pseudostates.

We substitute expansion (2) into the Schrödinger equation (1). Then apply the Hamiltonian operator on the wave function and use the semiclassical approximation [15]. Multiplying the result by $\psi_{\alpha'}^*(\boldsymbol{r}_\alpha) e^{-i\boldsymbol{q}_{\alpha'} \cdot \boldsymbol{\rho}_\alpha}$ $(\alpha' = 1, \dots, N)$ and integrating over $\boldsymbol{r}_\alpha$, and then multiplying by $\psi_{\beta'}^*(\boldsymbol{r}_\beta) e^{-i\boldsymbol{q}_{\beta'} \cdot \boldsymbol{\rho}_\beta}$ $(\beta' = 1, \dots, M)$ and integrating over $\boldsymbol{r}_\beta$, after some lengthy algebra we get the following set of first-order differential equations for the time-dependent coefficients

$$\begin{cases} i\dot{F}_{\alpha'} + i\sum_{\beta=1}^{M} \dot{G}_\beta \widetilde{K}_{\alpha'\beta} = \sum_{\alpha=1}^{N} F_\alpha D_{\alpha'\alpha} + \sum_{\beta=1}^{M} G_\beta \widetilde{Q}_{\alpha'\beta}, & \alpha' = 1,2,\dots,N, \\ i\sum_{\alpha=1}^{N} \dot{F}_\alpha K_{\beta'\alpha} + i\dot{G}_{\beta'} = \sum_{\alpha=1}^{N} F_\alpha Q_{\beta'\alpha} + \sum_{\beta=1}^{M} G_\beta \widetilde{D}_{\beta'\beta}, & \beta' = 1,2,\dots,M, \end{cases} \quad (3)$$

where dots over F_α and G_β denote time derivative. The system of equations is solved subject to the initial boundary condition specified as

$$\begin{cases} F_\alpha(-\infty, \boldsymbol{b}) = \delta_{\alpha 1}, & \alpha = 1,2,\dots,N, \\ G_\beta(-\infty, \boldsymbol{b}) = 0, & \beta = 1,2,\dots,M. \end{cases} \quad (4)$$

Overlap integrals $K_{\beta'\alpha}$, direct-scattering matrix elements $D_{\alpha'\alpha}$ and electron-transfer matrix elements $Q_{\beta'\alpha}$ are given in [6]. Notably, they coincide with the corresponding matrix elements from the conventional close-coupling theory. In addition, the set of equations (3) is also the same. This is despite us starting from the exact Schrödinger equation and using a different ansatz for the total wave function. However, a novel feature of our approach is that it does not use the concept of a so-called electron translation factor, required in the conventional approaches due to an inadequate representation of the electron-capture channels. Some six decades ago Bates and McCarroll [16] realised that the conventional expansion of the electronic part of the wave function does not satisfy the semiclassical time-dependent Schrödinger equation unless an electron translation factor is introduced to remedy the problem. The problem with this artificial electron translation factor is that there are infinitely many ways of choosing it [15]. Our ansatz correctly represents both direct scattering and rearrangement channels, and therefore does not lead to the aforementioned problem. Hence, there is no need to introduce any remedial factor.

3. Results

The CCC approach has been applied to calculate cross sections [4-6, 12-14] and stopping powers [20-23] for hadrons in various atomic and molecular targets in the energy range from 1 keV to 1 MeV. Recently, we have used the wave-packet implementation of the CCC (WP-CCC) to calculate differential cross sections in proton-hydrogen collisions [6]. In particular, doubly differential cross sections for ionisation gave excellent agreement with experimental measurements available at 75 keV. Figure 1 shows the WP-CCC results for angular differential cross sections for elastic scattering, $n = 2$ excitation and electron capture in proton-hydrogen collisions at 60 keV. Excellent agreement is observed with experimental measurements [17-19] for every process. See [6] for details of the calculations and more results.

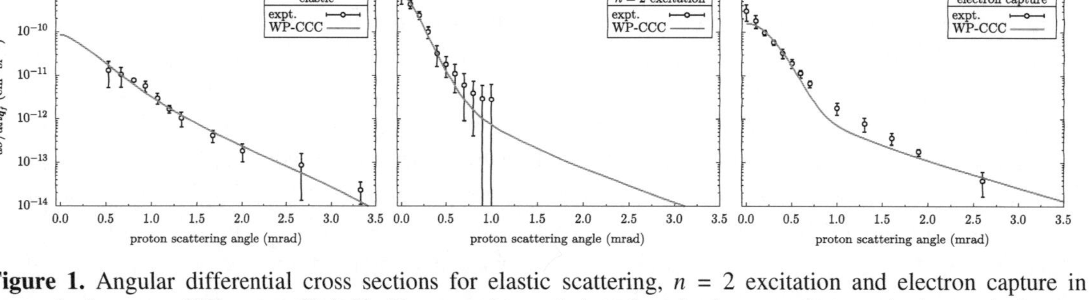

Figure 1. Angular differential cross sections for elastic scattering, $n = 2$ excitation and electron capture in proton-hydrogen collisions at 60 keV. The experimental data for elastic scattering, excitation and electron capture are from [17], [18] and [19], respectively.

4. Conclusion

We have developed a convergent close-coupling approach to hadron scattering on atomic and molecular targets applicable in the Bragg peak region. The approach starts from the exact fully quantum-mechanical three-body Schrödinger equation for the total scattering wave function. The scattering wave function is expanded using a two-center pseudostate basis that includes the continuum in a suitably discretized form. This allows one to take into account all underlying processes: elastic scattering, excitation, ionisation, and electron capture into bound and continuum states of the projectile. After using the semiclassical approximation we get a system of coupled differential equations for the transition probability amplitudes, well-known in conventional close-coupling methods based on the semiclassical time-dependent Schrödinger equation for the electronic part of the scattering wave function. However, a novel feature of our approach is that it does not use the concept of a so-called electron translation factor, required in other approaches due to an inadequate representation of the electron-capture channels. The convergent close-coupling approach is capable of providing the most detailed and accurate description of the processes taking place in hadron interactions with simple targets.

5. Acknowledgments

This work was supported by Australian Research Council.

6. References

[1] Bolst D *et al* 2017 *Nucl. Instr. Meth. Phys. Res. A* **869** 68-75
[2] Zeitlin C and La Tessa C 2016 *Front. Oncol.* **6** 65
[3] Belkic D 2010 *J. Math. Chem.* **47** 1366-419
[4] Avazbaev S K *et al* 2016 *Phys. Rev. A* **93** 022710
[5] Abdurakhmanov I B *et al* 2016 *Phys. Rev. A* **94** 022703
[6] Abdurakhmanov I B *et al* 2018 *Phys. Rev. A* **97** 032707
[7] Bray I, Abdurakhmanov I B *et al* 2017 *J. Phys. B: At. Mol. Opt. Phys.* **50** 202001
[8] Mukhamedzhanov A M *et al* 2014 *Phys. Rev. C* **90** 034604
[9] Fraser P R *et al* 2017 *Phys. Rev. C* **96** 014619
[10] Tursunov E M *et al* 2018 *Phys. Rev. C* **97** 035802
[11] Blokhintsev L D *et al* 2018 *Phys. Rev. C* **97** 024602
[12] Abdurakhmanov I B *et al* 2013 *Phys. Rev. Lett.* **111** 173201
[13] Abdurakhmanov I B *et al* 2015 *Phys. Rev. A* **91** 022712
[14] Abdurakhmanov I B *et al* 2017 *Phys. Rev. A* **96** 022702
[15] Bransden B H and McDowell M R C 1992 *Charge Exchange and the Theory of Ion-Atom Collisions* (Oxford: Clarendon)
[16] Bates D R and McCarroll R 1958 *Proc. R. Soc. A* **245** 175-83
[17] Rille E *et al* 1984 *Phys. Rev. A* **29** 521-4
[18] Park J T *et al* 1980 *Phys. Rev. A* **21** 751-58
[19] Martin P J *et al* 1981 *Phys. Rev. A* **23** 3357-60
[20] Bailey J J *et al* 2015 *Phys. Rev. A* **92** 022707
[21] Bailey J J *et al* 2015 *Phys. Rev. A* **92** 052711
[22] Bailey J J *et al* 2016 *Phys. Med.* **32** 1827-32
[23] Bailey J J *et al* 2017 *J. Phys. Conf. Ser.* **777** 012010

IOP Conf. Series: Journal of Physics: Conf. Series **1154** (2019) 012014 doi:10.1088/1742-6596/1154/1/012014

SABR in clinical trials: what quality assurance (QA) is required and how can it be done?

T Kron[1,2,3] and N Hardcastle[1,3]
[1]Peter MacCallum Cancer Centre, Melbourne, Australia
[2]Sir Peter MacCallum Cancer Institute, University of Melbourne
[3]Centre for Medical Radiation Physics, University of Wollongong, Australia

E-mail: tomas.kron@petermac.org

Abstract. Stereotactic Ablative Body Radiotherapy (SABR) has become widely used radiotherapy treatment modality for many indications ranging from early stage disease to an oligometastatic setting. All this combines the use of virtually all technological advances that characterise modern radiotherapy from small conformal fields to intensity modulation and explicit accounting for motion. However, given its novelty, clinical evidence for SABR is still emerging and many clinical trails include SABR as the intervention to define (eg in dose or fractionation defining studies) or to test against other treatment modalities. In any case, physical quality assurance (QA) must be an essential aspect of the trial to ensure that SABR is delivered accurately and in a comparable fashion in radiotherapy centres participating in the trial. The present paper tries to explore the balance between the QA cost and benefit for trial conduct in terms of patient numbers and timely completion of the trial.

1. Introduction

Stereotactic Ablative Body Radiotherapy (SABR) or Stereotactic Body Radiation Therapy (SBRT) as it is called in North America has quickly developed into a widely used treatment modality for a number of cancers [1-3]). However, high-level clinical evidence is only slowly emerging and there is general agreement that more clinical trials are required to define the intervention of SABR in terms of dose, fractionation and delivery technique as well as compare it with other treatment modalities and explore potential late effects [4]. Given the stark contrast in modalities such as surgery and SABR for early stage lung disease it is also difficult to recruit patients to these trials as equipoise may neither exist for an individual patient nor their caring clinician [5-8].

It is difficult to demand evidence for many technology-driven changes as has been discussed extensively in the context of IMRT and proton radiotherapy [9]. However, high-level evidence with clinical endpoints is desirable for the introduction of new technology and governments; regulatory authorities and health insurance companies usually demand it. In this context high-level evidence is defined as the outcomes of at least one randomised controlled trial. In practice, often not only the effectiveness of the new intervention is considered but also the resource requirements such as time and cost. A standard decision-making matrix is shown in figure 1. It is interesting to note that the quality assurance activities for clinical trials can help to identify some of the 'other factors' considered in the two blue diagonal fields. Robustness against unintended variations, complexity, training requirements and cost would be some of these factors.

IOP Conf. Series: Journal of Physics: Conf. Series **1154** (2019) 012014 doi:10.1088/1742-6596/1154/1/012014

Some additional issues are also worth consideration when looking for clinical evidence in a technology driven field like radiation oncology:

- Technology changes quickly and clinical endpoint may take a long time to be tested. As such, the results of a clinical trial may refer to old technology by the time they are mature.
- Implementation of technology and techniques may vary more amongst centres participating in a trial that tests new approaches for which no clinical guidelines exist.
- There is often a learning curve with new techniques; trials testing them may not reflect the optimum implementation at the early stage of a trial.

All this leads to an increased need to provide stringent quality assurance procedures for such trials. The current paper explores how this can be done in the context of SABR.

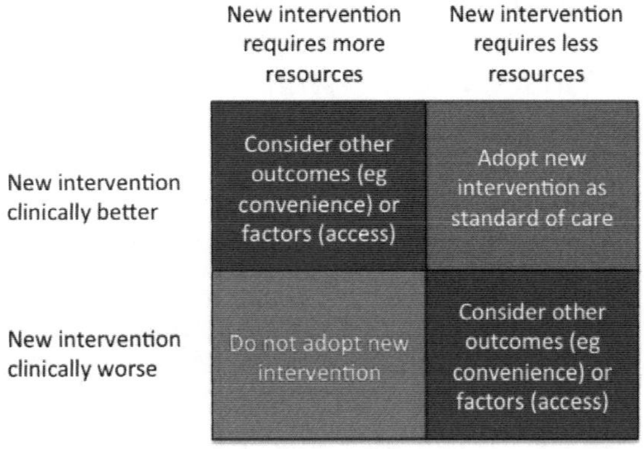

Figure 1. Typical scheme to introduce new technologies or techniques ('interventions').

2. Technical challenges for SABR

The concept of SABR is derived from intracranial stereotactic procedures where high radiation doses are given to small lesions in a very conformal fashion since more than 50 years using mostly rigid frames attached to the patient's skull to ensure accurate field localisation. However, compared to intracranial delivery, SABR is considerably more difficult as the body consists of heterogeneous, non-rigid materials, and the target is likely to move against the fixed reference frame typically used for image guidance [10, 11]. Figure 2 shows an illustration of typical SABR scenarios. Image guidance has largely replaced physical frames as the stereotactic 'frame of reference.

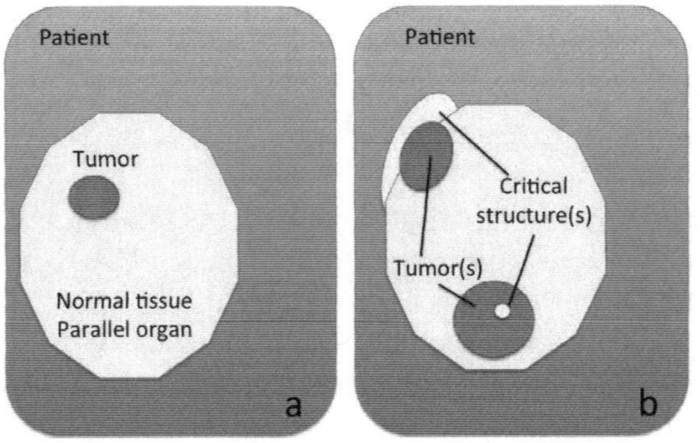

Figure 2. Schematic drawing of two SABR scenarios which may require different technical solutions: a) the original paradigm where a target is embedded in a large normal organ, typically a parallel organised radiobiological structure. Lung and liver are examples for this. b) a more complex SABR scenario consists of critical structures being either in close proximity or even enclosed by the target. Vertebral and prostate SABR treatments would be example for this.

Several clinical trials using scenario a in figure 2 are currently closed to accrual either because they are completed, or they failed to recruit. While the concept of SABR was pioneered in Scandinavia, [12, 13] it was the phase I trial by Timmerman et al performed in a single institution that inspired a paradigm shift in radiation oncology [14, 15]. Not surprisingly, the need for quality assurance in stereotactic body radiotherapy was identified almost immediately [16].

3. Physical QA for trials and their impact

The impact on clinical outcomes and thus clinical trial findings of quality assurance has been documented in several settings [17, 18]. In several circumstances the lack of adequate quality assurance may have prevented the trial from answering the trial question [19-21] or has affected clinical outcomes [22, 23]. This leads to the primary objective of trial QA to ensure that the trial is best positioned to answer the trial question. It is important to note that trial QA therefore does NOT ensure safe patient treatment in itself but aims to support the investigator in addressing the hypothesis of the trial. Figure 3 illustrates the resulting dilemma for a trial chair: optimising accrual by making it as easy as possible to contribute to the trial or tightening the quality criteria to standardise the trial and improve data quality.

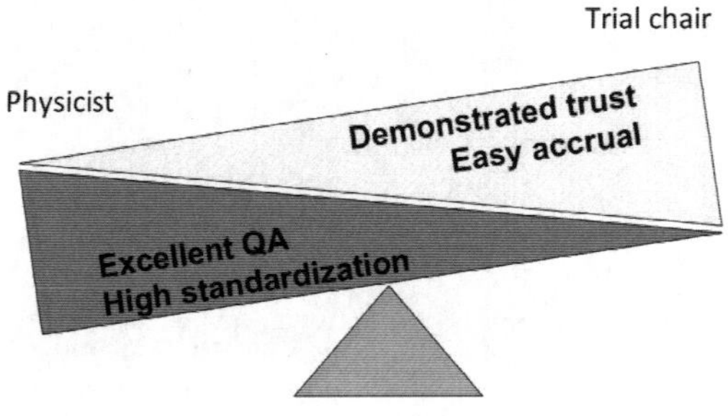

Figure 3. Illustration of the perceived trade-off between QA effort and trial accrual

A different consideration though is that high quality data generally reduces the quantity of data required to answer a question and thus can reduce the number of patients (and therefore the cost and time) required for a trial [20]. This has been elegantly demonstrated by Pettersen et al in 2008 [24]. When considering how few events determine the outcome of a trial this need for stringent quality control of data becomes obvious.

Trials that employ high-end technology such as SABR trials are particularly vulnerable to technological issues and careful consideration must be given to QA tools employed. In general trials use several tools to ensure quality data from a technology perspective [25]. Of particular relevance are:

- Credentialing of centres prior to being activated for accrual – this typically includes a benchmarking planning study and dosimetric audits
- Review of treatment plans for individual patients

The extreme hypofractionation in SABR relies on severe restrictions of dose to normal tissues utilising high-end technology and techniques. Their appropriate use requires generally stringent QA measures. Given that small field dosimetry is more complicated than dosimetry for conventional radiotherapy, dosimetric audits become particularly important and Clark et al provide a good summary of this in the recent literature [26]. Small field dosimetry was also identified as one issue in a recent study on credentialing for a SABR trial by our group [27]. Dosimetric audits for small fields are also offered by the IROC centre in Houston which also published a very useful compilation of small field output data [28]. The problem of dosimetry in small fields gets even more complicated if flattening filter free

(FFF) beams are used where spectral variations and high dose rate can complicate dosimetry further [29].

Furthermore, heterogeneities have posed issues for trial dosimetry [30]. It is interesting to note that retrospective analysis of the dose calculation algorithms in RTOG0236 led to an adjustment of the dose per fraction in the three fractions treatments from 20Gy to 18Gy to account for the lack of adequate inhomogeneity corrections in the dose calculation [31]. This has been shown to be problematic by our group and has also been identified also by the Australian Clinical Dosimetry Service (ACDS), a national dosimetric audit network [27, 32]. A number of dosimetric audits for SABR in lung cancer have attempted to address this [26, 27, 33]. More modern dose calculation algorithms such as Monte Carlo and use of the linear Boltzman equation introduce further complexity with the concept of dose to medium. Variations in dose reporting can lead to significant differences in delivered doses between centres that use dose to medium and those who use more traditional algorithms which report dose to water, in particular to bony targets [34, 35].

In the context of clinical trials dosimetric audits can be organised in different ways. They can be centrally organised such as in Australia [36] and North America [37, 38] or use a round robin approach where centres audit each other [39] The advantage of the centrally organised audits is that they are usually associated with a National Standard Laboratory, which ensures traceability of dose. Given the workload these centres also usually operate remotely using mail-out phantoms. Round Robins are more likely to include site visits that have the advantage that they can include education and other observations that are useful in the context of clinical trials [27, 40].

In the context of SABR also motion management and image guidance are important. Motion management audits are part of several IROC audits for liver and lung SABR [41, 42] and they have also been used in UK [43] and Australia [27, 44]. Similarly, image guidance is an essential aspect of trial QA and several methods have been reported in the literature [45-50].

4. Conclusion

There is no perfect quality assurance program for SABR, in particular given the wide variety and increasing number of targets treated. However, there is consensus that participation in clinical trials for SABR does require some independent quality assurance activities to ensure technology and techniques are used in a consistent way across all participating centres. As most clinical trials to date are considering SABR of lung lesions, heterogeneities, motion and small fields are key aspects of the checks. However, given the move of SABR applications to targets very close to critical structures as illustrated in figure 2, also the steepness of dose fall off and location of the dose gradient and as such image guidance are becoming increasingly important. As SABR trials start to involve more than one lesion in lung, liver and brain it will be interesting to see how audits for these scenarios will be developed in the future.

5. Acknowledgments

We would like to acknowledge the financial support of the Gross Foundation for our SABR program.

6. References

[1] Andratschke N *et al* 2018 *BMC Cancer* **18** 283
[2] Daly M E *et al* 2013 *J. Thorac. Oncol.* **8** 202-7
[3] Guckenberger M *et al* 2013 *J. Thorac. Oncol.* **8** 1050-8
[4] Treasure T *et al* 2015 *Lancet Oncol.* **16** 597-8
[5] Chang J Y *et al* 2015 *Lancet Oncol.* **16** 630-7
[6] Bahig H *et al* 2017 *J. Thorac. Dis.* 9 953-6
[7] White A *et al* 2016 *J. Thorac. Dis.* 8 S399-405
[8] Laba J M *et al* 2017 *Curr. Oncol.* 24 e524-30
[9] Duchesne G M *et al* 2015 *J. Med. Imaging Radiat. Oncol.* **59** 363-70
10. Dawson L A *et al* 2007 *J. Clin. Oncol.* **25** 938-46
[11] Kron T 2012 *Cancer Forum* **36**

[12] Blomgren H *et al* 1995 *Acta Oncol.* **34** 861-70
[13] Lax I *et al* 1994 *Acta Oncol.* **33** 677-83
[14] Timmerman R *et al* 2003 *Chest* **124** 1946-55
[15] Timmerman R *et al* 2003 *Technol. Cancer Res. Treat.* 2153-60
[16] Timmerman R *et al* 2006 *Acta. Oncol.* **45** 779-86
[17] Bentzen S M 1998 *Radiother. Oncol.* **46** 5-18
[18] Peters L J *et al* 2010 *J. Clin. Oncol.* **28** 2996-3001
[19] Rischin D *et al* 2010 *J. Clin. Oncol.* **28** 2989-95
[20] Perez C A *et al* 1984 *Int. J. Radiat. Oncol. Biol. Phys.* **10** Suppl 1:119-25
[21] Valley J F *et al* 1998 *Radiother. Oncol.* **47** 37-44
[22] Fairchild A *et al* 2013 *Int. J. Radiat. Oncol. Biol. Phys.* **87** 246-60
[23] Ohri N *et al* 2013 *J. Natl. Cancer Inst.* **105** 387-93
[24] Pettersen M N *et al* 2008 *Radiother. Oncol.* **86** 195-9
[25] Kron T *et al* 2013 *J. Phys. Conf. Ser.* **444** 012014
[26] Clark C H *et al* 2017 *Phys. Med.* **44** 171-6
[27] Kron T *et al* 2018 *Br. J. Radiol.* **91** (1085) 20170737
[28] Followill D S *et al* 2012 *J. Appl. Clin. Med. Phys.* **13** 3962
[29] Xiao Y *et al* 2015 *J. Appl. Clin. Med. Phys.* **16** 5219
[30] Latifi K *et al* 2014 *Int. J. Radiat. Oncol. Biol. Phys.* **88** 1108-13
[31] Xiao Y *et al* 2009 *Int. J. Radiat. Oncol. Biol. Phys.* **73** 1235-42
[32] Dunn L *et al* 2015 *Phys. Med.* **31** 435-41
[33] Distefano G *et al* 2017 *Radiother. Oncol.* **122** 406-10
[34] Andreo P 2015 *Phys. Med. Biol.* **60** 309-37
[35] Moiseenko V *et al* 2013 *Phys. Med. Biol.* **58** 7107-16
[36] Lye J *et al* 2014 *Med. Phys.* **41** 032102
[37] Ibbott G S *et al* 2013 *Front. Oncol.* **3** 311
[38] Ibbott G S *et al* 2008 *Int. J. Radiat. Oncol. Biol. Phys.* **71** (1 Suppl): S71-5
[39] Drzymala R E *et al* 2015 *Med. Phys.* **42** 6745-56
[40] Healy B *et al* 2013 *Med. Phys.* **40** 071706
[41] Taylor P A *et al* 2016 *Int. J. Radiat. Oncol. Biol. Phys.* **95** 242-8
[42] Vinogradskiy Y Y *et al* 2009 *Med. Phys.* **36** 3438-47
[43] Palmer A L *et al* 2017 *Radiother. Oncol.* **125** 453-8
[44] Hardcastle N *et al* 2014 *Australas. Phys. Eng. Sci. Med.* **37** 45-52
[45] Kry S F *et al* 2012 *J. Appl. Clin. Med. Phys.* **13** 3939
[46] Middleton M *et al* 2011 *Int. J. Radiat. Oncol. Biol. Phys.* **81** 1576-81
[47] Healy B J *et al* 2003 *Med. Phys.* **30** 2282-91
[48] Vial P *et al* 2006 *Phys. Med. Biol.* **51** 5517-38
[49] Vial P *et al* 2008 *Med. Phys.* **35** 4362-74
[50] Gustafsson H *et al* 2009 *Med. Phys.* **36** 5665-74

IOP Conf. Series: Journal of Physics: Conf. Series **1154** (2019) 012015 doi:10.1088/1742-6596/1154/1/012015

Ionoacoustics for range monitoring of proton therapy

S Lehrack, W Assmann and K Parodi
Ludwig-Maximilians-Universität München, Faculty of Physics, Department of Medical Physics, Am Coulombwall 1, 85748, Garching b. München, Germany

E-mail: Katia.Parodi@lmu.de

Abstract. Proton beams exhibit favourable interaction properties for highly conformal tumour dose delivery with better sparing of surrounding healthy tissue and critical structures in comparison to widely established photon beams. However, full clinical exploitation of these advantages is still hampered by treatment uncertainties, particularly related to the stopping position of protons in tissue, determining the so called Bragg peak, where protons release most of their energy. Hence, several approaches are currently investigated to enable an in-vivo, ideally real-time verification of the proton beam range in the patient. While most of the considered techniques aim at exploiting secondary emissions from nuclear interactions, interest is being regained for alternative, cost-effective solutions exploiting acoustic emissions originating from ion interaction in tissue. This contribution will thus address the basics and recent experiments of this so called "ionoacoustics" for pre-clinical and clinical beam energies.

1. Introduction

Proton therapy is an emerging treatment modality exploiting the favourable interaction properties of swift ions in matter. Along with advanced scanning delivery methods, proton therapy can enable very good targeting of the dose to the tumour with excellent sparing of surrounding healthy tissue and critical structures compared to widely established external photon beam therapy. However, full clinical exploitation of the promised advantages in the clinical practice is still hampered by the main issue of range uncertainties, i.e., uncertainties in the position of the well localized dose deposition of individual pencil beams in the so-called Bragg peak (BP). To this end, most of the currently investigated techniques aim at exploiting secondary emissions from nuclear interactions less obviously correlated to the dose deposition, such as positron emission tomography and prompt gamma imaging [1]. Here, we present an alternative approach exploiting acoustic emissions originating from ion interaction in tissue, so called "ionoacoustics" [2]. This method has recently been revisited at two medical facilities in USA [3] and France [4]. After reviewing the basic principles, initial experience at the two facilities along with systematic measurements at low proton beam energies are presented and compared.

2. Material and methods

When penetrating a medium, protons mainly lose energy in electronic collisions, resulting in localized heating and a thermal expansion, which generates thermoacoustic emissions detectable with acoustic transducers. However, only spatial and temporal gradients can produce a pressure wave in such adiabatic heating. These gradients are generated temporally by pulsed beams with steep rise times, and spatially by locally confined energy deposition as in the Bragg peak. Under these conditions different spatial source locations can be identified for pencil-like beams depending on the detector position and proton range. For short ranges (e.g. 4.1 mm for 20 MeV protons as shown in figure 1) a so called direct signal expanding spherically around the BP can be measured, along with a signal generated at the entrance in water and a reflection of the direct signal at the water/air interface [1]. By placing a suitable transducer, e.g., axially and distal to the BP, the range of protons can be directly calculated from the time difference between the entrance or the reflection signal and the direct signal knowing the water temperature (for accurate speed of sound determination) [1]. For higher proton energies and longer ranges, an additional signal can be seen from the cylindrical ultrasound wave expanding only laterally from the beam axis. A full theoretical description can be found for example in [5].

To better understand the complex dependence of the ionoacoustic signal strength and shape on the spatial and temporal distribution of the dose deposition, we performed a large series of measurements of ionoacoustic emissions from pulsed proton beams (~ns for pre-clinical energies, ~ns for clinical energies) in water, varying possible beam parameters (e.g, pulse length/intensity) that influence the signal amplitude. Since the properties of the used hydrophone in sensitivity and size are also important, comparative measurements between currently used detectors are shown. The presented data in water are from a 20 MeV proton beam studied at the MLL Tandem accelerator in Garching near Munich, Germany [1], a ≥ 200 MeV clinical beam from a superconducting synchrocyclotron at the Centre Antoine Lacassagne (CAL) in Nice, France [3], and a comparative measurement conducted at the Roberts Proton Therapy Centre located at the university of Pennsylvania (UPENN) in Philadelphia, USA, with a 230 MeV proton beam.

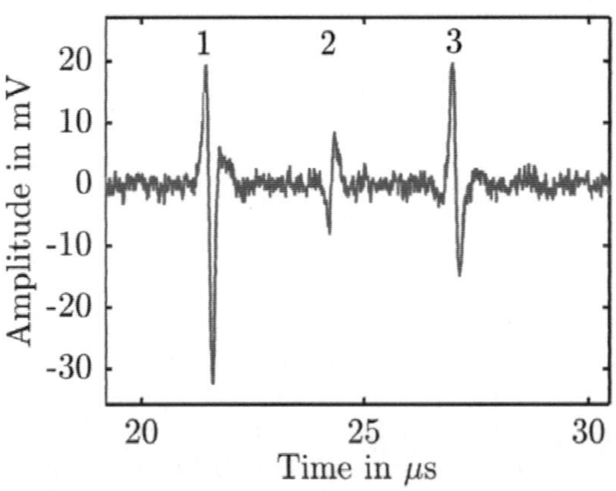

Figure 1. Exemplary ionoacoustic signal from a pulsed 20 MeV proton beam stopped in water [1]. Different signal parts correspond to different source locations: 1) the direct signal from the BP, 2) the signal generated at the entrance point of the protons into water, 3) a reflection of the direct signal.

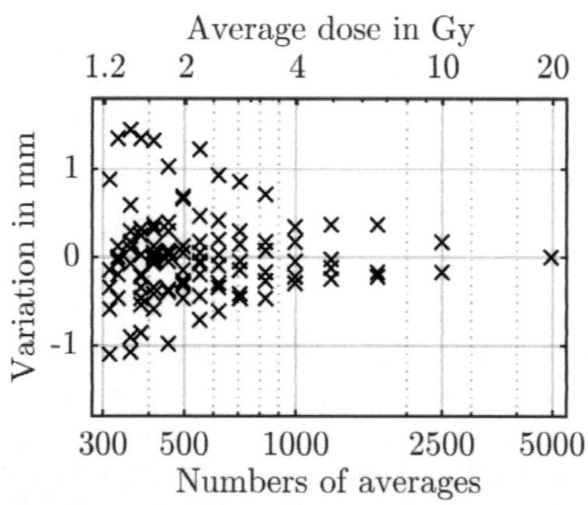

Figure 2. Variation of pulse averages for a 200 MeV proton beam at CAL [4, https://doi.org/10.1088/1361-6560/aa81f8]. The averaged dose was calculated for a cylinder surrounding the Bragg peak, where the dimensions were defined by experimental spot size measurements with radiochromic films. (© Institute of Physics and Engineering in Medicine. Reproduced by permission of IOP Publishing. All rights reserved)

3. Results and discussion

The linear dependency of the signal amplitude with the deposited energy was confirmed with a 20 MeV proton beam and a pulse width of 473 ns, varying the particle number per pulse from ca. 10^5 to 10^7 protons/pulse. A polynomial fit of the data showed a linear relation with a detection limit for this setup of about 10^5 particles/pulse with averaging of 16 events, corresponding to a dose deposition of about 1.6 Gy in the Bragg peak region [1]. A similar study was then conducted with 200 MeV proton and a 3.7 μs pulse width at CAL. Since the signal amplitude at these higher beam energies is significantly lower, averages up to 1000 pulses were required. In a study with 5000 consecutive single pulses, the amount of averages was successively reduced and the resulting precision in time-of-flight (TOF) range determination is shown in figure 3, demonstrating that an averaged dose of 2.8 Gy is required for a range retrieval precision of ±1 mm [4]. Here, doses were averaged in a cylindrical volume around the Bragg peak with dimensions deduced from radiochromic film measurements.

The frequency spectrum of the ionoacoustic signal is, amongst others, determined by the heating spatial distribution, hence depending on the initial proton kinetic energy and energy dependent lateral and range straggling. While for the 20 MeV proton beams the FWHM of the Bragg peak was about 300 μm, providing a central frequency of ~2.5 MHz, the FWHM of a 230 MeV proton beam increases to ~28 mm, thus decreasing the central frequency of the ionoacoustic signal to ~10 kHz. Results presented in [3] and figure 2 show that range determination with > 200 MeV proton beams with sub-millimetre resolution is possible and hence not limited by the low central frequency of the ionoacoustic signal. The challenge arising from these broad heat distributions is the choice of a proper detector matched to the signal frequency. Additionally, with longer proton ranges the TOF for reflection of the direct signal increases (figure 1) and the amplitude of window and reflection signals is significantly lower than for the direct signal, thus challenging their detection.

Another very important parameter is the pulse width. Two important time scales are commonly given for two regimes: thermal and stress confinement. The first one is the time limit for adiabatic heating, usually in the order of milliseconds. For pulse length below the thermal confinement, heat diffusion can be neglected. Stress confinement is fulfilled, if the pulse width is shorter than the flight time of the generated acoustic signal through a characteristic volume, in this case the Bragg peak. When stress confinement is fulfilled, the temporal profile can be considered as a delta spike excitation providing optimal condition for pressure generation with highest amplitudes. In this case, the resulting acoustic signal is mainly defined by the spatial heating distribution. The stress confinement for a 230 MeV proton beam is a pulse width of 18.5 μs. At the isochronous cyclotron at UPENN, this stress confinement was achieved with a pulse width of 17 μs [3] by pulsing the arc current. However, better conditions are provided by the synchrocyclotron at CAL, with inherently pulsed proton beams with a width of 3.7 μs [4]. Along with the higher pulse intensity of 2 pC per pulse, this synchrocyclotron provides optimal conditions for ionoacoustic measurements, translating in better precision and accuracy for proton range determination. Here, we measured for the first time the ionoacoustic signal with a bidirectional, flat Cetacean C305x hydrophone, simultaneously using a scintillator to define the start of the TOF measurement from the proton entering the water phantom to the arrival time of the highest pressure wave carrying the signal from the Bragg peak [4]. With a precise levelling of the hydrophone to the water surface, this distance was subtracted from the total depth of the hydrophone, yielding the range of the proton in water. From the comparison with ionization chamber (IC) depth measurements done directly following the ionoacoustic measurements with the same energy settings, as well as from previous commissioning measurements, a sub-millimeter range accuracy and precision could be confirmed for sufficient signal strength, i.e., deposited dose (figures 2 and 3).

Finally, comparative measurements were conducted at UPENN, comparing our C305x hydrophone with their omnidirectional calibrated Brüel & Kjær 8105, used for the results reported in [3]. To this end, the hydrophones were placed axially to an artificially pulsed 230 MeV proton beam in a water phantom, as in [3]. The position of the hydrophones was fixed and the proton beam range was reduced by introducing additional plates of solid water into the beam path. The increased distance of the Bragg peak from the two hydrophones is shown in dependence of the introduced additional absorbers in figure 4. The intended shift is shown by the black dashed line. In general, both hydrophones were able to follow the intended BP shift, with a slight underestimation for the 8105 model. However, this latter hydrophone was found less sensitive to noise and distortion, so that its precision is better compared to the fluctuation observed in the results from the C305x.

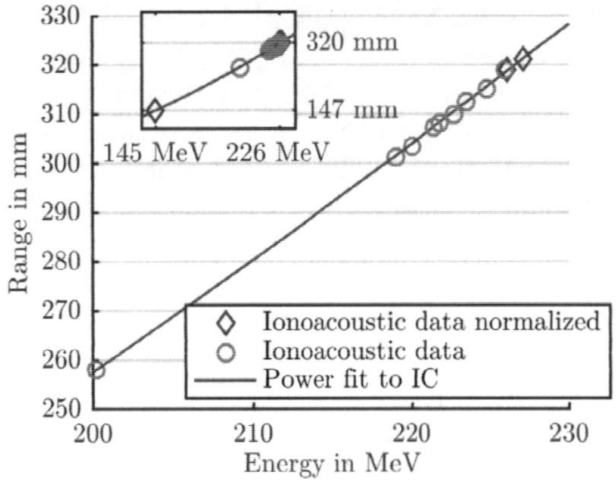

Figure 3. Compilation of ionoacoustic-deduced ranges in water, compared to a fit of reference IC measurements at CAL ([4], https://doi.org/10.1088/1361-6560/aa81f8. (© Institute of Physics and Engineering in Medicine. Reproduced by permission of IOP Publishing. All rights reserved).

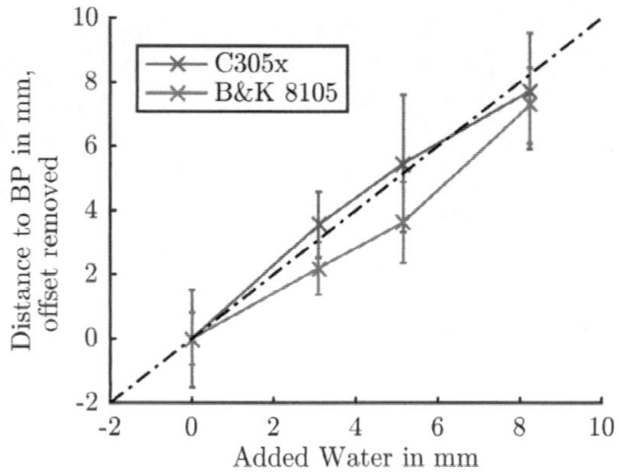

Figure 4. Comparative measurement of relative range shifts induced by additional plates of solid water, measured with two different hydrophones at UPENN. The dashed black line indicates the intended shift of the Bragg peak.

4. Conclusions

Recent systematic TOF measurements in water revealed the potential of (sub)millimetre accuracy and precision of ionoacoustics for range monitoring, provided that sufficient signal is collected, e.g., by averaging over more proton pulses for total doses up to 10 Gy [3,4]. While the considered low beam energies of ~20 MeV generate high frequency acoustic waves detectable with typical ultrasound transducers operating in the MHz range, clinical beam energies delivered by intrinsically or artificially pulsed (synchro)cyclotrons demand the usage of broadband, low-frequency (kHz) hydrophones. Although millimetre range monitoring based on ionoacoustics still demands relatively high fraction doses of a few Gy and is challenged by increased absorption and scattering in patient-like heterogeneous tissue compared to water, the reported results encourage ongoing efforts of several groups to improve the detection efficiency and to account for tissue heterogeneities with proper simulation and image reconstruction methods [6,7]. Ultimately, ionoacoustics could offer a compact and cost-effective modality for real-time in-vivo verification of the beam range (and possibly also anatomy with additional co-registered ultrasound imaging), at least for suitable clinical tumour indications of easy sonic access such as prostate, breast, liver [1].

5. Acknowledgement

We thank K. C. Jones and S. Avery for sharing results collected with their hydrophone and amplifier. The authors acknowledge the support of several collaborators from the Munich Military University (G. Dollinger), the Technical University Munich (V. Ntziachristos, S. Kellnberger and co-workers), the Centre Antoine Lacassagne (J. Herault, M. Vidal and co-workers), S. Reinhardt, P. Thirolf as well as several BSc and MSc students at LMU. This work was funded by the German Research Foundation, Cluster of Excellence Munich Center for Advanced Photonics (MAP).

6. References

[1] Parodi K and Polf J 2018 *Med. Phys.* (in press)
[2] Assmann W *et al* 2015 *Med. Phys.* **42** 567
[3] Jones K C *et al* 2015 *Med. Phys.* **42** 7090
[4] Lehrack S *et al* 2017 *Phys. Med. Biol.* **62** L20
[5] Jones K C *et al* 2016 *Med. Phys.* **43** 5213

IOP Conf. Series: Journal of Physics: Conf. Series **1154** (2019) 012015 doi:10.1088/1742-6596/1154/1/012015

[6] Jones K C *et al* 2018 *Phys. Med. Biol.* **63** 025018
[7] Patch S K *et al* 2018 *Med. Phys.* **45** 783

IOP Conf. Series: Journal of Physics: Conf. Series **1154** (2019) 012016 doi:10.1088/1742-6596/1154/1/012016

Recent developments at the University of Pennsylvania's Roberts Proton Therapy Center

R L Maughan, S Avery, E S Diffenderfer, D Dolney and B-K K Teo
Department of Radiation Oncology, University of Pennsylvania, 3400 Civic Boulevard, Philadelphia, PA 19085, USA

E-mail: Richard.Maughan@uphs.upenn.edu

Abstract. The Roberts Proton Therapy Center treated its first patient on Australia Day 2010 with a double scattered proton beam. In 2012 the first patient was treated with pencil beam scanning and by that time patients were being treated in five treatment rooms. By March 2013 we were treating with pencil beam scanning (PBS) in three treatment rooms and double scattering in the other two. We currently treat about 100 patients per day in 4 treatment rooms, while one of our double scattering beam nozzles is being replaced with PBS. We treat 15 to 20 pediatric patients per day, many under general anesthetic. We treat a wide variety of indications in adult patients and all patients are registered on some form of clinical research protocol. We were the first proton center to have an operational CBCT system (September 2014) and Small Animal Radiation Research Platform (SARRP) on one of the beam lines which can be used for both x-ray and proton irradiations to support our radiobiology research program. Areas of physics related research described here include prompt gamma range monitoring, proto-acoustic range verification and the development of micromegas detectors for range verification through proton radiography and CT.

1. Introduction

Recent developments at the Roberts Proton Therapy Center (RPTC) include innovations in clinical application, medical physics and radiobiology research. The RPTC is fully integrated into the Department of Radiation Oncology and incorporates an Ion Beam Applications Proteus Plus system (IBA SA, Louvain-la-Neuve, Belgium), with four gantry rooms and one fixed beam room. When the purchase agreement was signed in June 2006, there was no commercially available pencil beam scanning (PBS) system for proton therapy. PBS was included as a research and development project within the contract. To mitigate the risk associated with PBS uncertainty, all four gantry rooms were supplied with double scattering (DS) and uniform scanning (US) and only two were provided with additional PBS capability. The fixed beam room has a PBS only beam line. This resulted in a very complex multi-modality facility, however, opting for this complex system proved to be a good choice because delays in deployment of PBS did not delay full clinical operation in all rooms.

Once PBS was operational in three rooms it became clear that this was the modality of choice for most patients because of the superior dose distributions achievable. Therefore, the facility operated with three PBS and two DS treatment rooms, until a DS room closed down for upgrade to PBS in late 2016. This room resumed clinical treatments in March of 2018. Eventually planned upgrade of the remaining DS room will create a PBS only facility. This continual upgrade philosophy also saw the use of the first gantry mounted proton cone beam computed tomography (CBCT) unit in September 2014; three PBS gantries are equipped with this technology.

2. Clinical Operation

Table 1 shows the total number of patients (adult and pediatric), categorized by treatment site, treated at the RPTC, since January 2010 and projected through until the end of the 2018 financial year which starts on July 1st, 2017 and ends on June 30th, 2018. Note the decrease in the percentage of urological (prostate) cases treated from 35% in FY2011 to 17% in FY 2018. Also notable is the significant increase in the numbers GI, head and neck and thoracic cases treated, starting around FY2013, as the number of available PBS rooms increased by adding two PBS gantry rooms, which allowed for greater flexibility in treatment planning and improved dose distributions.

Table 1. Total number of patients and disease sites treated between January 2010 and July 2018

Disease Site	Financial Year								Total
	2011	2012	2013	2014	2015	2016	2017	2018	
Breast	1	0	6	13	30	44	28	31	**153**
Endocrine	13	21	17	26	27	25	22	14	**165**
Gastro-intestinal	14	48	93	136	117	116	120	86	**730**
Gynecology	3	4	6	16	14	10	3	9	**65**
Head & Neck	10	16	33	66	75	97	95	96	**488**
Hematology	4	17	23	51	50	48	60	58	**311**
Metastatic	3	2	11	9	16	18	32	24	**115**
Missing	13	21	46	5	2	1	4	14	**106**
Musculoskeletal	29	33	37	56	65	56	56	57	**389**
Neurological	56	76	112	132	119	149	164	165	**973**
Other	1	4	2	5	3	3	3	3	**24**
Skin	0	0	2	2	6	7	4	7	**28**
Thoracic	13	57	67	103	122	142	114	127	**745**
Urology	105	178	161	156	117	111	129	111	**1068**
Total	**265**	**477**	**614**	**772**	**763**	**827**	**832**	**964**	**5359**
The above totals include pediatric cases for which the annual treatment numbers are as follows;									
Pediatric cases	55	96	119	134	141	162	139	163	**1009**

3. Research Developments

3.1. Range verification

There are a number of projects addressing the issue of range uncertainty; these include prompt gamma emission studies, protoacoustics and proton radiography.

3.1.1. Prompt gamma emission. When protons lose energy in passing through the patient a portion of that energy is loss in elastic and inelastic nuclear interactions [1]. The inelastic interactions can leave the target nuclei (typically ^{12}C, ^{14}N and ^{16}O) in an excited state as shown in figure 1 or can produce new isotopes. The excited states de-excite through emitting gamma rays which generally have very short half-life (typically in the µs-ps range). The induced isotopes may also have short half-lives but many have longer ones of several minutes or more. This project has been in collaboration with IBA a prototype IBA prompt gamma system has been clinically tested.

The prototype uses a gamma camera with a knife-edge collimator to detect the gamma signal emitted along the path of the proton beam during PBS therapy. The distal edge of the individual pencil beam Bragg peaks may be detected. In a clinical trial the measured Bragg peak positions in a patient were compared with simulated Bragg peak profiles from the treatment planning system. The clinical subject was a brain patient treated with three fields; a lateral, an oblique and a vertex field. Prompt gamma emission imaging was performed on 6 of the 30 treatments over a 14 day period. The measured average range shifts for the lateral, oblique and vertex fields were -0.8±1.3 mm, 1.7±0.7 mm and -0.4±0.9 mm, respectively [2]. Typically, the position of the Bragg peak may be determined with an accuracy of 2mm.

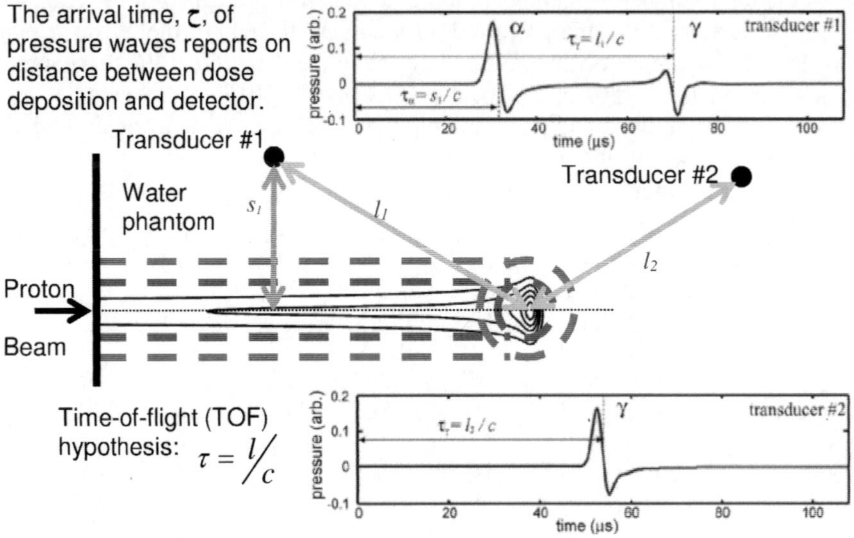

Figure 1. Schematic representation of the prompt gamma emission process.

3.1.2 Protoacoustics. An alternative and novel approach to range verification is to detect the acoustic pressure wave generated when protons slow down and stop. The first experimental measurements of this phenomenon by Sulak *et al* in 1979 [3], but recently there has been renewed interest in the technique, which has stimulated research at the RPTC. The principle relies on using multiple detectors to determine the "time-of-flight" of the pressure wave from its origin (proton path and Bragg peak) to the detector. This knowledge combined with the known velocity of sound in the phantom medium (c) is used to determine the position of the Bragg peak as illustrated schematically in figure 2.

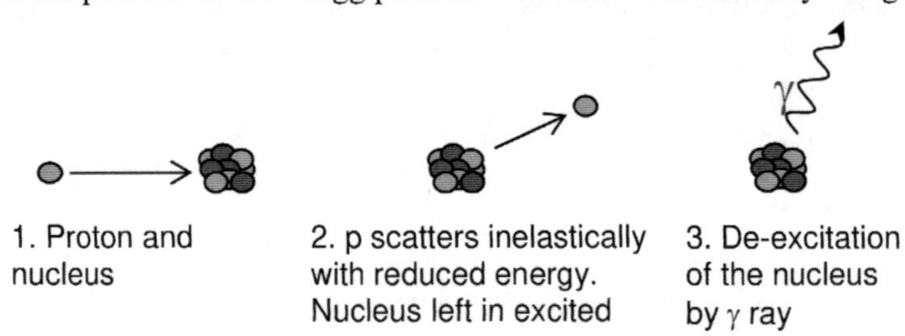

Figure 2. Schematic representation of the protoacoustic technique for range verification.

One problem is obtaining a sufficiently short proton pulse to provide adequate signal for detection. Calculations show that to efficiently generate protoacoustic signal the proton pulse width should ideally be less than 10μs. Synchrotrons and linear accelerators may satisfy this requirement but it is problematical for isochronous cyclotrons without modification, although synchrocyclotrons with their pulsed output produce pulses in this range. Another problem will be the need for accurate knowledge of the acoustic properties of the different tissues of the body; simulations addressing this issue have been published by [4].

3.1.3. Proton radiography. Micromegas detectors have been developed in collaboration with the University Physics and Astronomy Department [5]. These detectors can have multiple applications but most recently they have been applied to proton radiography. Pixelated detectors have been built and spatial resolutions of 0.3 mm has been demonstrated measuring the position of step functions in a phantom at image doses of 2 cGy. Water equivalent thickness (WET) can be measured with an accuracy of 60 μm. At 0.2 cGy image dose the resolution degrades to 0.9 mm. A multilayer device can be used to determine residual WET and reconstruct a proton radiograph. Initial measurements have been made using a single layer Micromegas detector and varying the incident beam energy.

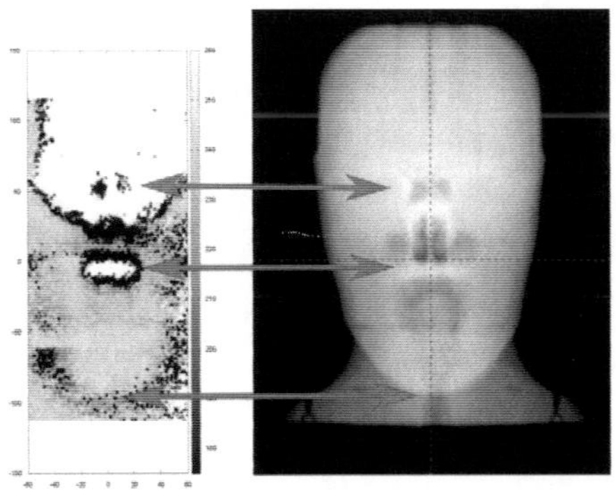

Figure 3. A proton image of a head phantom obtained by varying the beam energy and detecting the WET using a single layer Micromegas detector. The proton radiograph (on the left) is compared to a DRR generated from a CT image of the head phantom. These early results demonstrate the feasibility of using the Micromegas technology in this application.

3.2. *SARRP for radiobiology research*
A dedicated research room at the end of the clinical beam line has recently been commissioned. This room is equipped with two research beam lines. A small animal radiation research platform (SARRP), incorporating x-ray imaging and x-ray beam delivery, has been installed at the end of one beam line enabling the image guide irradiation of small animals with the proton beam.

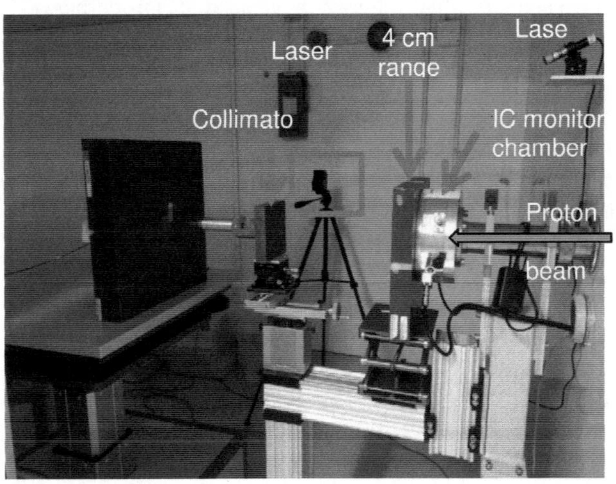

Figure 4. The proton research beam line during commissioning. The main components of the beam line visible: the monitor ionization chamber, the range shifter and the collimator. The SARRP now occupies the space where the test measurement bench is situated in this photograph. The SARRP is mounted on rails allowing to translate laterally out of the beam path allowing for other experimentally set ups to be installed.

4. Conclusions
Some of the more significant medical physics research projects currently undertaken at the University of Pennsylvania's Roberts Proton Therapy Center have been briefly described. It represents a small fraction of the wide variety of clinical, medical physics and radiobiology research performed at this institution using the proton beam.

5. References
[1] Gorjiara T *et al* 2012 *Med. Phys.* **39** 7071-9
[2] Xie Y *et al* 2017 *Int. J.Radiation Oncol. Biol. Phys.* **99** 210-18
[3] Sulak L *et al* 1979 *Nucl. Instr. Methods.* **161** 203-17
[4] Jones K C *et al* 2018 *Phys. Med. Biol.* **63** 025018
[5] Hollebeek R *et al* 2012 *Phys. Res. Int.* **2012** Article ID 714782

IOP Conf. Series: Journal of Physics: Conf. Series **1154** (2019) 012017 doi:10.1088/1742-6596/1154/1/012017

A FPGA-based software for microdosimetric data processing

D Mazzucconi[1,2], M Bonfanti[1,2], D Bortot[1,2], S Agosteo[1,2], A Pola[1,2], S Pasquato[1,2] and A Fazzi[1,2]
[1]Politecnico di Milano, Energy Department, via La Masa 34, 20156 Milano, Italy
[2]INFN, Sezione di Milano, via Celoria 16, 20133 Milano, Italy

E-mail: davide.mazzucconi@polimi.it

Abstract. The tissue equivalent proportional counter (TEPC) is the most accurate device for measuring microdosimetric properties of particle beams. Since microdosimetric quantities span over several decades, the electronic and acquisition chain should meet specific requirements. In order to cover the wide dynamic range of the signals generated by the TEPC, the output signal from the preamplifier is fed in parallel to three linear amplifiers which shape and amplify the signal with different gains. Very low-energy deposition events are filtered in the high-gain stage, and high-energy deposition events are processed in the low-gain stage. A new system with high acquisition performance and compact hardware was developed for this purpose. The analog-to-digital conversion is performed by a commercial acquisition system that includes a FPGA. Thanks to the FPGA a parallel high-speed acquisition on three channels can be performed. The software merges signals together from the three electronic chains and computes a real time microdosimetric spectrum giving a prompt information about the irradiation field. This acquisition system, which performs analog-to-digital conversion and signal processing at a sampling rate up to 15 MS/s, was tested by irradiating a TEPC with an Am-Be fast neutron field, an intense quasi-monoenergetic neutron beam and a 62 MeV/u helium ion beam.

1. Introduction

In the last decades, radiation therapy with ion beams has been spreading worldwide for treating cancer. With respect to conventional photon radiation therapy, better dose conformation to the target and increased sparing of the healthy tissue surrounding the tumour can be achieved with ion beams, due to the different physics of interaction of charged particles into matter. Moreover, higher LET (linear energy transfer) radiation, like carbon ions, exhibits a larger number of double strand breaks in the DNA and a higher RBE in the Bragg peak than photons and protons and is, thus, a good candidate for the treatment of radio-resistant tumours [1]. Since ionization generated in ion therapy is markedly localized and non-uniform against depth, significant variations on the radiation quality and consequently differences in the biological effectiveness across the Bragg curve are present [2]. Nevertheless, the present radiation-treatment planning procedures are based on the measurement of the absorbed dose, which is a macroscopic and average quantity that does not take into account neither the stochastics of particle interactions in the target volume nor the track structure of ionizing charged particles, which is crucial for the initiation of the radiation damage [3]. The standard dosimetric approach can be integrated with methodologies and instruments provided by microdosimetry, which aims at measuring the statistical fluctuations of the local energy imparted at the micrometric level [4].

IOP Conf. Series: Journal of Physics: Conf. Series **1154** (2019) 012017 doi:10.1088/1742-6596/1154/1/012017

The tissue equivalent proportional counter (TEPC) is the most accurate device for measuring the microdosimetric properties of a particle beam. Since microdosimetric quantities (i.e. specific energy and lineal energy) may span over several decades, the electronic and acquisition chains of such detectors should meet further requirements with respect to the conventional ones. In order to cover the wide dynamic range of the signals generated by the TEPC and to ensure a good resolution throughout this range, the output signal from the preamplifier is generally fed in parallel to three linear shaping amplifiers. The acquisition chain should be capable of processing and merging the signals coming from the three amplifiers. A new system with high acquisition performance, in terms of real time calculations, and compact hardware was developed for this purpose.

2. Materials and methods

Common TEPCs operate in pulse mode and record each single ionization event occurred in the sensitive volume. The interaction of an ionizing particle within the gas cavity is translated into a voltage signal distinguishable from the electron noise by exploiting a proper electronic acquisition chain. Since ionization events occurred in the sensitive volume of the proportional counter span over a very wide range, a low-noise charge-sensitive preamplifier with a dynamic range higher than 4 decades is required. Moreover, the preamplifier should be connected as close as possible to the anode wire in order to maximize the signal to noise ratio. Signals from the preamplifier are processed by a further amplification stage constituted by three linear amplifiers in parallel which shape and amplify with different gains (usually set to obtain a relative ratio of 1, 10 and 100) for covering the wide range of produced signals. In this way, very low-energy deposition events (mainly due to electrons) are filtered in the high-gain stage, whereas very high-energy deposition events (mainly due to recoil and low-energy ions), which necessarily saturate in the high-gain stage, are processed by the low-gain stage.

A typical TEPC acquisition chain is shown in Figure 1.

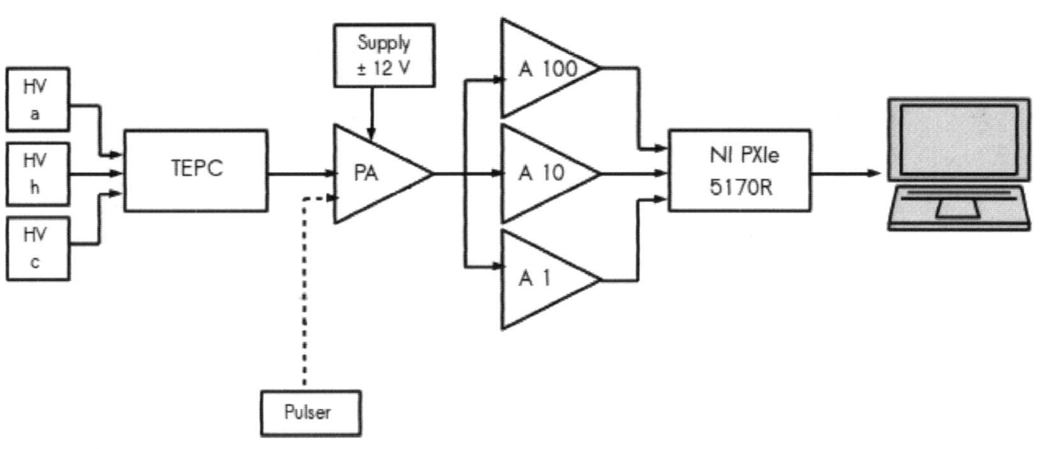

Figure 1. Block diagram of the TEPC whole electronic and acquisition chain.

The analog-to-digital conversion is performed by a commercial acquisition system produced by National Instruments, the 4 channels 14 Bit-Oscilloscope NI PXIe-5170R, which is a configurable digitizer including a user-programmable FPGA (Field Programmable Gate Array) module for the on-board signal processing. The FPGA module is an integrated circuit that contains a matrix of reconfigurable gate-array logic circuitry that is programmed via software. When an FPGA is configured, the internal circuitry is connected in a way that creates a hardware implementation of the software application. In this way, the FPGA is capable of performing high speed parallel computations on the acquired data. The FPGA circuitry is programmed with the LabVIEW FPGA module, which belongs to the high-level LabVIEW graphical programming environment. The FPGA module also contain a built-in FIFO (First In First Out) transfer and memory read/write functions for storing data in the FPGA application. The FIFO transfer is the connecting bridge between the FPGA and the elaboration software, which gives the microdosimetric spectrum. The FIFO memory is the core of this implementation, which allows in the meanwhile a high-speed acquisition through the FPGA module and the computations needed for producing the microdosimetric spectrum. High speed is reached by decoupling the acquisition part (performed by the FPGA that stores the pulses into the memory) from the microdosimetric elaboration which is carried out by the host PC. The development of this software required several efforts in order to maintain high computation speed, since the complex calculations needed to provide the final microdosimetric spectrum affect unavoidably the final efficiency of the whole process.

3. Results

Thanks to the FPGA module a parallel high-speed acquisition on the three channels can be performed. Moreover, the implemented software can merge the three electronic chains and compute a real time microdosimetric spectrum. The developed software has a graphical interface where the user can set the acquisition parameters (e.g. sampling rate and acquisition thresholds) and visualize the microdosimetric spectrum.

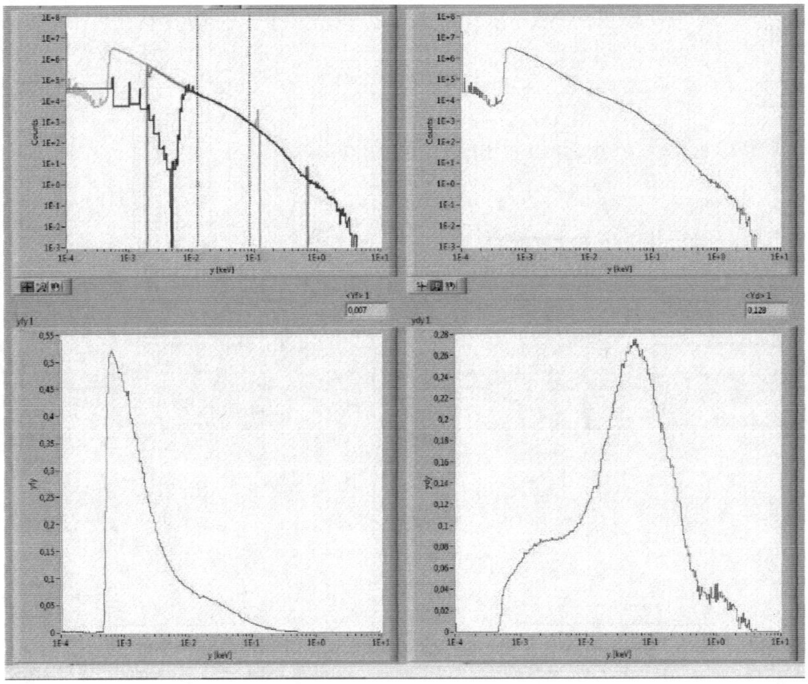

Figure 2. Software user interface in which the microdosimetric elaboration is shown. Top-left: spectra from each channel. Top-right: joint and rebinned spectrum. Bottom-left: frequency distribution yf(y). Bottom-right: dose distribution yd(y). Note that the neutron microdosimetric spectrum shown in this figure is not calibrated against lineal energy.

The FPGA software can perform analog-to-digital conversion and digital signal processing at a sampling rate up to 15 MS/s per each channel (the ADC speed is 45 MS/s). This sampling rate is enough for obtaining a good digitisation of pulses even when the shaping time of the amplifiers is set to 250 ns. This low value is mandatory for minimizing signal pile-up when the TEPC is irradiated with intense fields (i.e. accelerator-based fields). During acquisition, the software can plot a real time microdosimetric spectrum and compute microdosimetric mean quantities (e.g. dose-averaged lineal energy) in order to obtain a prompt information about the irradiation field.

An example of real time computation on the acquired data is shown in Figure 2, where a portion of the user interface is reported. The spectra shown in the figure are obtained by irradiating the avalanche-confinement TEPC previously described [5-7] with an Am-Be neutron source at the Politecnico di Milano fast-neutron irradiation facility. The figure shows the different steps of the elaboration starting from the merging of the outputs of the three electronic chains (top-left) to the final microdosimetric spectrum (bottom-right). Further irradiations with quasi-monoenergetic neutron beams and 62 MeV/u helium ion beam were performed with this system. In particular, measurements against the helium beam showed that a real time information about the microdosimetric distribution is really useful and suitable for a first evaluation of the radiation quality of a therapeutic hadron beam.

4. Conclusions

The new FPGA-based hardware and software allow to reach a significant high acquisition speed required when measuring with intense radiation fields, since very short shaping times are mandatory in order to avoid pulse pile-up. The new acquisition software was tested by irradiating a TEPC with an Am-Be fast neutron field, an intense quasi-monoenergetic neutron beam and a 62 MeV/u helium ion beam and showed the capability of producing real time information (microdosimetric spectrum and mean quantities) about the irradiation field. Further optimization on the FPGA architecture is foreseen in order to achieve an improvement on the computation performances and to allow the implementation of other kind of microdosimeters (e.g. silicon based) in the elaboration software.

5. Acknowledgments

This work was supported by the Italian National Institute for Nuclear Physics - INFN – Scientific Commission V in the framework of the MITRA (Microdosimetry and TRAck structure) and NADIR (biologically relevant NAnoDosimetry of Ionizing Radiation) projects.

6. References

[1] Durante M and Loeffler J S 2010 *Nat. Rev. Clin. Oncol.* **7** 37-43
[2] Gorjiara T *et al* 2012 *Med. Phys.* **39** 7071-9
[3] Weyrather K and Kraft G 2004 *Radioth. Oncol.* **73** S161-9
[4] International Commission on Radiation Units and Measurements 1983 *Microdosimetry ICRU Report* **36**
[5] Bortot D *et al* 2017 *Radiat. Meas.* **106** 531-7
[6] Bortot D *et al* 2017 *Radiat. Meas.* **103** 1-12
[7] Bortot D *et al* 2017 *Radiat. Prot. Dosim.* **180** 172-6

IOP Conf. Series: Journal of Physics: Conf. Series **1154** (2019) 012018 doi:10.1088/1742-6596/1154/1/012018

Head and neck plan quality: where we are and does it matter?

B Nelms[1], J Caudell[2], D Gintz[2] and V Feygelman[2]
[1]Canis Lupus LLC, Merrimac, Wisconsin 53561, USA
[2]Moffitt Cancer Center, Tampa, Florida 33612, USA

E-mail: vladimir.feygelman@moffitt.org

Abstract. Radiation treatment planning for the Head and Neck (HN) cancers is challenging and complex, and here we investigate various aspects of HN plan quality. We present the results of an international plan quality study, in which a web-based platform (ProKnow) was used to analyse a total of 238 plans submitted from 34 countries, all for the same dataset, anatomy, and quantified, objective plan scoring algorithm. All treatment planning systems (TPS) studied were able to produce high quality plans: 6 of 6 TPS models had scores in the top 25%, and 4 of those in the top 10%. However, all TPS and modalities also showed substantial variability in plan quality across all planners, suggesting needs for training and/or automation. Concerning current automation solutions, both commercially available approaches (AutoPlanning and Knowledge-Based Planning) are so far unable to outperform experienced human planners when tight target dose coverage and homogeneity are required. Target dose homogeneity is not to be discounted, as it may be linked with the rate of complications such as reactive gastrostomy tube placement after oropharynx chemoradiation.

1. Introduction

In this work, we review the state of affairs on a variety of topics pertinent to the quality of Head and Neck (HN) radiation treatment planning. We cover (i) the results of a recent, international study that collected and measured metric-based plan quality as well as dosimetric accuracy (measured vs. calculated) for a challenging case; (ii) early experiences with planning automation, including published results from various groups including ours; and (iii) potential link between target dose homogeneity and the rate of clinical complications.

2. Treatment planning study

2.1. Methods

A web-based platform (ProKnow, http://www.proknowsystems.com) was used to run the study. A published HN patient CT from MPPG 5a [1] was provided with pre-contoured targets and organs-at-risk (OAR), and a set of dosimetric metrics and corresponding objectives. The plan included a primary target and bilateral elective neck node CTV treated using simultaneous integrated boost to three dose levels. Uploaded plans were scored in real-time and with full transparency using an objective, composite scoring formalism [2]. A total of 21 metrics were extracted, scored function-wise, and summed for each plan, resulting in the composite Plan Quality Metric score (PQM). Target dose coverage and homogeneity made up 63% of the possible total, with the remainder accounting for OAR sparing. The relative weight of the different dose objectives, while clinically realistic, is by necessity somewhat subjective, as it reflects preferences of the study team. However, once the weights are fixed, the rest of the scoring is applied objectively to all participants, who knew the details up front and were left with the task of maximizing their score as they deemed fit after the minimum constraints and objectives were met. The participants were free to choose the treatment planning system (TPS), treatment modality, beam energy, and planning techniques, provided they were clinically viable and practical. Interactive histograms and scatterplots of all metrics are available to the public via the ProKnow website.

In addition to generating histograms of PQM distributions, statistical comparisons between different sub-groups were performed. The median PQMs were compared with non-parametric Kruskal-Wallis ANOVA followed by Dunn's multiple comparisons test (GraphPad Software, La Jolla California USA) between the following treatment modalities: C-arm linac VMAT and IMRT, helical tomotherapy (HTT), and protons. All treatment planning systems were included. For IMRT, segmented and dynamic plans were scored together, as were scattered and scanned proton beams. P-value below 0.05 was considered an indication of statistical significance. In addition, other descriptive statistics per TPS were recorded, such as the minimum and maximum PQM and the standard deviations.

For standard linacs, the median VMAT vs. IMRT PQM were additionally pair-wise compared per TPS by a Mann-Whitney non-parametric test when a reasonable number of IMRT plans were available (i.e. Eclipse and Monaco). Next, the median photon PQM scores were compared with the Kruskal-Wallis ANOVA test between five different TPS.

The median number of monitor units and/or estimated beam-on time were compared between the photon modalities: VMAT, IMRT, and HTT. Spearman correlation coefficient $\boldsymbol{r}$ was computed between VMAT and IMRT PQM and corresponding total MU. Similarly, a possible correlation was explored between the VMAT, IMRT, and HTT PQM and estimated beam-on time.

Finally, the "pre-treatment" dosimetric QA results were requested and analysed in terms of the method employed and γ-analysis passing rates, with the 3% (global)/3 mm/10% low dose threshold, 3% (global)/2mm/10%, and 2% (local)/2 mm/20% criteria combinations.

2.2. *Results*

A total of 238 plans were submitted from 34 countries. 39% were from the USA. In terms of treatment modalities, VMAT made up 73% of the submissions, followed by fixed gantry IMRT (18%), tomotherapy (6%), and protons (3%).

The participation by TPS and overall distribution of the PQM are shown in figures 1 and 2, respectively.

The overall median VMAT (N=173) score was 135.6 vs. 130.1 for IMRT (N=43). The overall maximum PQM of 146.9 was achieved with a 3-arc VMAT plan generated by the Eclipse TPS, requiring only 642 total MU. The ANOVA comparison between different modalities showed no statistically significant differences. No significant differences were also found between the median VMAT vs. IMRT PQM for the individual TPS. The results of non-parametric ANOVA comparisons of median PQM for five different TPS with the sufficiently high number of entries are presented in Table 1 along with some descriptive statistics. All TPS were able to produce high quality plans: 6 of 6 models had scores in the top 25%, and 4 of those in the top 10%. Statistical significance was only demonstrated for Monaco's PQMs, which were lower pair-wise comparisons than all other systems except HTT, reaching high significance ($p<0.005$) against Pinnacle and RayStation. The results of comparisons between the efficiency (total MU or time) of different photon modalities are presented in table 2. The median number of MU for VMAT was statistically significantly lower than for IMRT ($p<0.0001$), and the same holds for beam-on time. VMAT beam-on time was significantly less than HTT ($p<0.0001$), and so was IMRT time ($p<0.0001$).

IOP Conf. Series: Journal of Physics: Conf. Series **1154** (2019) 012018 doi:10.1088/1742-6596/1154/1/012018

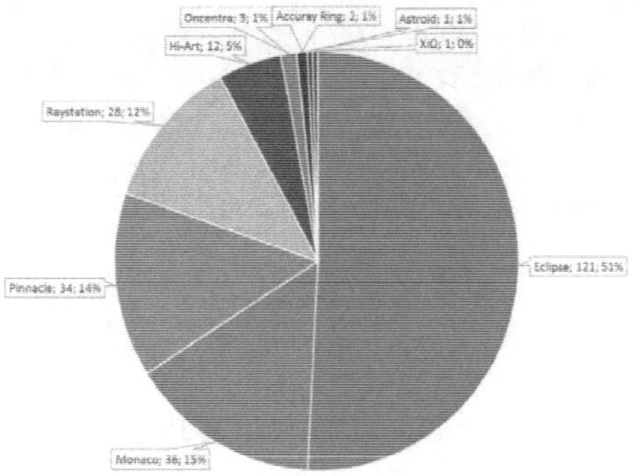

Figure 1. Participation by TPS.

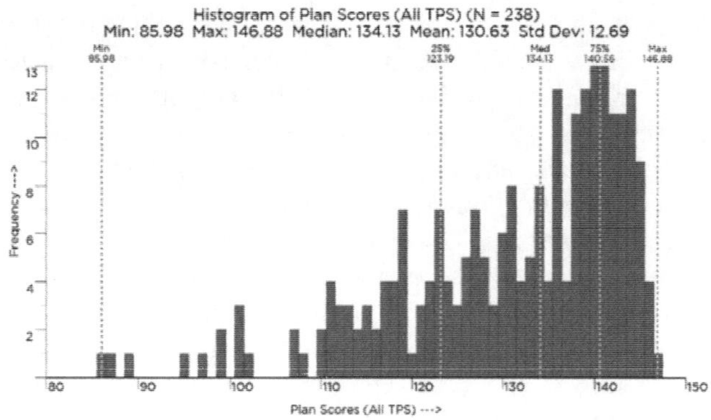

Figure 2. Overall PQM score distribution.

Table 1. Descriptive statistics of photon PQM for different TPS and p-values for the median score pair-wise tests between TPS.

	N	Min PQM	Median PQM	Max PQM	SD	p-value			
						HTT	Monaco	Pinnacle	RayStation
Eclipse	120	86.0	133.6	146.9	12.5	>0.9	***0.03***	0.25	>0.9
HTT	15	100.8	131.1	145.8	12.7		0.7	>0.9	>0.9
Monaco	36	99.1	125.9	141.5	11.8			***0.0003***	***0.0045***
Pinnacle	34	114.5	138.8	144.6	7.8				>0.9
RayStation	26	88.9	138.5	145.8	13.5				

Table 2. Monitor units and estimated beam-on times for different photon modalities

Plan efficiency metric	MU		Time (min)		
Statistic	VMAT	IMRT	VMAT	IMRT	HTT
N	172	42	153	19	15
Min	308	433	1.7	2.6	5.1
Median	596	977	3.0	5.0	9.18
Max	1616	3159	4.8	8.1	27.1
SD	198	696	0.6	1.3	6.1

No correlation was found between the PQM and MU for either VMAT or IMRT (max $r < 0.46$).

Over 95% of the reported (77) QA tests used either a true (65%) or collapsed single angle (31%) composite measurements. The results support setting aggressive accuracy goals in terms of TPS algorithm commissioning. The median 2% (local normalization)/2mm gamma-analysis passing rate was 90%, with the top quartile of performers showing passing rates over 95%.

The most important finding in this section is that all TPS and modalities showed substantial variability in plan quality distribution, suggesting needs for planner training and/or automation. The latter is the subject of the following discussion.

3. Attempts at automation

3.1. AutoPlanning

AutoPlanning (AP) is offered in Pinnacle TPS (Philips). It attempts to mimic the steps of a human treatment planner while performing iterative optimization based on dynamically created regions of interest. Technically, it does not require prior knowledge of similar plans. When quantitatively compared to manually designed plans with the emphasis on target dose coverage and homogeneity, similar to Section 2, AP on average scored lower than the human planner [3]. While excelling in sparing OARs, the software lacks reliable tools to improve target coverage parameters to the levels required by our institution, such as <105% single voxel hot spot.

3.2. Knowledge-Based Planning
Knowledge-Based Planning (KBP) is a feature in Eclipse TPS (Varian). It tries to predict achievable dose-volume histograms based on the library of the previously accepted plans that have similar relative geometry of the targets and OARs. We are not aware of any published study demonstrating that a library of previous high-quality HN plans would ensure equally high-quality output. Studies such as [4] used too loose target coverage/homogeneity criteria to draw conclusions in that regard.

3.3. Possible clinical implications
PQM scoring in the HN plan challenge and our AP evaluation [3] emphasized target dose homogeneity. To our knowledge, no study directly connecting IMRT/VMAT target dose homogeneity with complication rates exists. However, there is some published evidence that the rates of reactive gastrostomy tube placement after oropharynx chemoradiation are substantially lower when a strict 105% hot spot limit is enforced [5], compared to a study [6] presumably based on less strict RTOG-type [7] objectives that allow, as a minor deviation, up to 117% target dose to 1 cc of tissue. This observation is particularly important given that the results similar to [6] are used to justify more expensive radiotherapy modalities in HN cancer [8], without fully exhausting the capabilities of photon treatments with conventional linacs [9].

4. Conclusions
Dosimetric quality of HN plans remains highly variable and operator-dependent. The long-term solution for this problem may be automation. However, in the realm of commercially available TPS, automated solutions are so far not competitive with humans when tight target dose coverage and homogeneity are required. There is some published evidence, admittedly far short of unequivocal, that such requirements may be contributing to reduction in complication rates.

5. References
[1] Smilowitz J B *et al* 2015 *J. Appl. Clin. Med. Phys.* **16** 14-34
[2] Nelms B E *et al* 2012 *Pract. Radiat. Oncol.* **2** 296-305
[3] Gintz D *et al* 2016 *J. Appl. Clin. Med. Phys.* **17** 331-46
[4] Tol J P 2015 *Int. J. Radiat. Oncol. Biol. Phys.* **91** 612-20
[5] Strom T *et al* 2013 *JAMA Otolaryngology - Head & Neck Surgery* **139** 1242-6
[6] Setton J *et al* 2015 *Cancer* **121** 294-301
[7] RTOG 1016. https://www.rtog.org/ClinicalTrials/ProtocolTable/StudyDetails.aspx?study=1016
[8] Blanchard P *et al* 2016 *Radiother. Oncol.* **120** 48-55
[9] Baldock C 2009 *J. Phys.: Conf. Ser.* **164** 012002

IOP Conf. Series: Journal of Physics: Conf. Series **1154** (2019) 012019 doi:10.1088/1742-6596/1154/1/012019

Progress of radiotherapy technology with HIMAC

K Noda
National Institute of Radiological Sciences, QST, 4-9-1 Anagawa, Inage-ku, Chiba 263-8555, Japan

E-mail: noda.koji@qst.go.jp

Abstract. NIRS has conducted a carbon-ion radiotherapy with HIMAC since 1994. For the progress of the carbon-ion radiotherapy technology, NIRS has developed beam-delivery technologies such as a broad-beam, a pencil-beam 3D-scanning and a superconducting rotating gantry. These technologies have been applied for both the static and moving tumor treatments. Further, NIRS just starts "Quantum Scalpel" project. In this project, a multi-ions irradiation for the LET-painting is studied to obtain the higher treatment efficacy. An ultra-compact heavy-ion radiotherapy machine, with much lower cost, has also been being developed by applying the superconducting and laser technologies for widespread use of the heavy-ion radiotherapy in the world.

1. Introduction

Heavy-ion (HI) beams are very suitable for deeply-seated cancer treatment not only owing to their high dose localization around the Bragg peak, but also owing to the high biological effect there [1]. NIRS, therefore, constructed HIMAC as the world's first heavy-ion accelerator facility dedicated to medical use [2]. NIRS has conducted a carbon-ion radiotherapy (RT) with HIMAC since 1994, which results in accumulating the number of patient treated of more than 11,000. The carbon-ion RT has so far proven to be significantly effective against radio-resistivity tumours such as a bone cancer and a malignant melanoma, while keeping the quality of life without any serious side effects. It has also reduced the number of treatment fractions in treatments of common cancers, which leads to a shorter course treatment than the low LET RT. NIRS, thus, proposed a standard carbon-ion RT facility [3] to boost the carbon-ion RT in Japan, with emphasis being placed on a compact version so as to reduce a construction cost. As a result of the development, both the facility size and costs were to be around one-third compared with those of HIMAC. The fruits of this work were boar in a pilot facility in the Gunma University that has been successfully conducted since 2010.

Since 2006, on the other hand, NIRS has been engaged in a "New Treatment Research Project" for further development of the HIMAC RT, such as the "adaptive cancer radiotherapy". NIRS, thus, has developed a fast 3D scanning with a pencil beam. The new treatment research facility in order to verify the technology proposed through the clinical study was constructed. The first patient was treated in May 2011 in a fixed beam-delivery room. A compact carbon-ion rotating gantry with superconducting (SC) technology [4] has been utilized for static-tumor treatments since April 2018.

NIRS has just proposed a new development plan, which called it "Quantum Scalpel" project. Objectives of this project are to obtain a higher treatment efficacy of the HI RT and to realize an ultra-compact RT machine with much lower cost. As an LET-painting technique for controlling a biological effect considered the relative biological effectiveness (RBE) and the oxygen enhancement ratio (OER) on a tumor, the multi-ions irradiation scheme has been being studied [5]. For ultra-compact machine, further, the quantum scalpel is to employ a laser-acceleration injector, a SC synchrotron and a SC rotating gantry.

This paper reviews the development and introduces a future plan of HIMAC.

2. Development of pencil-beam 3D scanning and rotating gantry

2.1. Pencil-beam 3D scanning

The pencil-beam 3D scanning is very suitable for the adaptive cancer RT, because this method requires neither the compensator nor patient collimators taking long manufacturing time. A pencil-beam 3D scanning is a beam delivery method to paint the dose distribution with a small beam and narrow Bragg peak. The pencil beam is laterally scanned (lateral scanning) to form a lateral dose distribution with orthogonal scanning dipole magnets and is then longitudinally scanned with a certain depth (slice) by either an energy absorber or by stepwise energy change from the accelerator.

2.1.1 Lateral scanning

Appling the pencil-beam 3D scanning to a moving-tumor treatment is difficult, because beam positions prescribed are moving. NIRS, therefore, has developed a phase-controlled rescanning (PCR) [6-8], based on the fast 3D scanning, for both the static and moving tumor treatments. The PCR method utilizes the fact that the tumor movement is close to "zero" on average during one respiration period. When a rescanning irradiation of one slice is completed within one respiratory-gated period, the PCR can obtain a uniform dose distribution even under irradiation of the moving tumors. The PCR method requires mainly two technologies: 1) Intensity-modulation technique for a constant irradiation time on each slice with a different cross section [9, 10] and 2) Fast 3D-scanning with pencil beam for completing several-times rescanning within a tolerable time [6].

2.1.2 Depth scanning

In the 3D scanning, the irradiated depth should be longitudinally changed, which called it depth scanning. In the first step, energy absorbers were used for the depth scanning. However, a variable-energy operation by accelerator itself has great advantages over that by the energy absorber; keeping the spot size small and suppressing secondary neutron yield. The multiple-energy operation, therefore, has been developed applying the extended flat-top operation of synchrotron. At first step of variable energy operation, an eleven-step energy operation from 430 to 140 MeV/n was developed [11, 12], In this scheme, range of more than 3 cm is changed by the energy change with the synchrotron, while that of less than 3 cm by thin energy absorbers. As the second step, the 201-step energy operation, which can change the energy ranging from 430 to 56 MeV/n, has been developed [13]. The energy change in one step corresponds to a range shift of 2 mm.

2.2. Rotating gantry

A SC rotating gantry with the fast 3D-scanning, has been developed [2] for the more accurate and shorter-course treatment. The gantry is designed to deliver carbon-ions with the energy of up to 430 MeV/n to the iso-center with irradiation angles of over ±180 degrees, and the maximum irradiation field of 20 ×20 cm^2. The SC rotating gantry consists of ten combined-function SC magnets, a scanner and two pairs of beam profile-monitors and steering magnets. The gantry length and the diameter are around 13 and 11 m, respectively. Since April 2018, the SC rotating gantry has been in practical use for a static-tumor treatment and been carried out as the clinical study for moving tumor.

3. Quantum scalpel project

NIRS has proposed the "Quantum Scalpel" project to obtain a higher treatment efficacy and to realize an ultra-compact RT machine with much lower cost for widespread use in the world.

IOP Conf. Series: Journal of Physics: Conf. Series **1154** (2019) 012019 doi:10.1088/1742-6596/1154/1/012019

3.1. Multi-ions irradiation
In the HIMAC treatment with a carbon-ion beam, a biological-dose distribution, described by (absorbed dose) ×(RBE), has been delivered to be constant over a planning target volume (PTV), while the RBE and OER distributions have not yet been controlled. It is noted that the RBE increases with increasing the LET (< ~200 keV/μm) while the OER decreases with increasing the LET (~20 keV/μm < LET < ~200 keV/μm). In this case, the LET distribution is shown in Figure 1 (a). This irradiation scheme gives the higher biological effect around the PTV boundary which is close to a normal tissue, because the Bragg-peak regions from various directions distributes around the PTV boundary. On the other hand, the lower one appears in the central region of the tumor. When a hypoxic state is in the central region of the tumor, a carbon-ion beam might not be effective due to highly resistivity against radiation. As one of the RBE-control methods, it was proposed that the RBE can be uniformly distributed across the PTV with two opposed carbon-ion fields [14]. NIRS has proposed the multi-ions irradiation to control the biological effect over the PTV. As shown in Figure 2 (b), the central region of the PTV is covered with the higher biological effect by mainly oxygen-ion, the PTV boundary with the lower one with mainly the helium-ion and the other region with middle one by mainly carbon-ion. It is noted that one can obtain a uniform biological-dose distribution even in this case. Combining a rotating gantry with 3D scanning, a multi-ions RT with the IMPT [15] will bring much higher treatment efficacy.

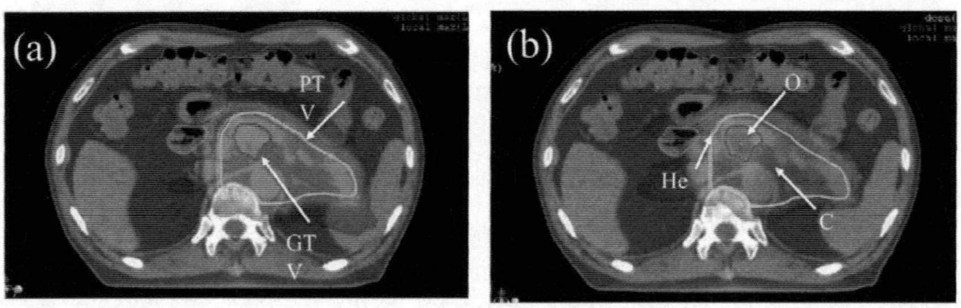

Figure 1. LET distribution, (a) with only carbon-ion, (b) with helium, carbon and oxygen-ions.

3.2. Ultra-compact HI RT machine
NIRS has studied a conceptual design of an ultra-compact HI-RT machine with the superconducting and laser technologies. The size of the machine is 10 m ×20 m, corresponding to two photon RT rooms. This machine consists of a laser-acceleration injector with 4 MeV/u, a SC synchrotron ring with around 7 m in diameter and a SC rotating gantry. In a conceptual design, the SC synchrotron can accelerate ions from 56 to 430 MeV/u with an injection energy of 4 MeV/u. The dipole field of the main magnet is designed to be 4 T at maximum and the field-change rate of around 1 T/s. The SC rotating gantry can be significantly downsized by a higher magnetic field of around 5 T and a compact scanner. As a result of a beam test of the compact scanner, the irradiation field size of 24 cm ×24 cm in 3.5 m of the distance from the entrance of the scanner to the iso-center, while the present scanner requires 9 m to obtain the same field size. A layout of the quantum scalpel is shown in Figure 2.

4. Summary
Almost 1 million and 14 million persons are diagnosed with cancer every year in Japan and in the world, respectively. In such a situation, NIRS has progressed the carbon-ion RT technologies. NIRS developed especially a standard carbon-ion RT facility in Japan, which bore fruit in a pilot facility in Gunma University. Following on from the pilot facility, three additional facilities have been successfully conducted and one facility has been being constructed in Japan. After ten years, further, it is forecast that the cancer patient number will continue to rise to 22 million persons every year in the world. Therefore, NIRS starts the "Quantum Scalpel" project, which will have a higher treatment efficacy with much lower cost to boost the heavy-ions RT in the world.

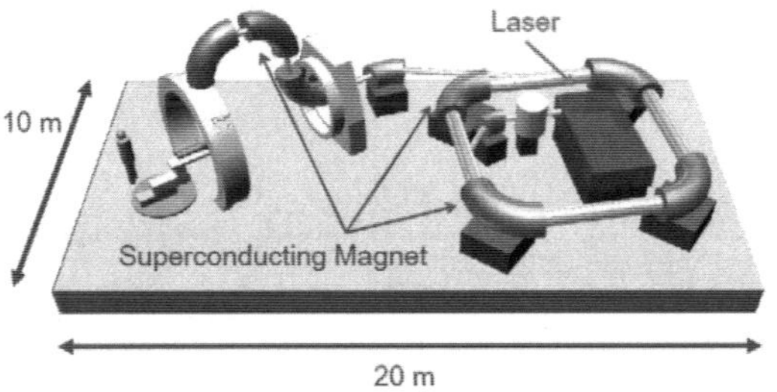

Figure 2. Layout of Quantum scalpel.

5. References

[1] Gorjiara T *et al* 2012 *Med. Phys.* **39** 7071-9
[2] Hirao Y *et al* 1992 *Nucl. Phys. A* **538** 541c-50c
[3] Noda K *et al* 2007 *J. Radiat, Res.* **48** A43-A54
[4] Iwata Y *et al* 2012 *Phy. Rev. ST-AB* **15** 044701
[5] Inaniwa T *et al* 2017 *Phys. Med. Biol.* **62** 5180-97
[6] Furukawa T *et al* 2007 *Med. Phys.* **34** 1085-97
[7] Furukawa T *et al* 2010 *Med. Phys.* **37** 4874-9
[8] Furukawa T *et al* 2011 *Med. Phys.* **37** 5672-82
[9] Sato S *et al* 2007 *Nucl. Instr. Meth. A* **574** 226
[10] Mizushima K *et al* 2011 *Nucl. Instrum. Meth. B* **269** 2915
[11] Iwata Y *et al* 2010 *Nucl. Intsrum. Meth A* **624** 33
[12] Inaniwa T *et al* 2012 *Med. Phys.* **39** 2820-5
[13] Mizushima K *et al* 2014 *Nucl. Instrum. Meth. B* **331** 243-7
[14] Böhlen T *et al* 2012 *Phys. Med. Biol.* **57** 7983-8004
[15] Lomax A J 1999 *Phys. Med. Biol.* **44** 185-205

IOP Conf. Series: Journal of Physics: Conf. Series **1154** (2019) 012020 doi:10.1088/1742-6596/1154/1/012020

Simulation of cosmic radiation spectra for personal microdosimetry at the International Space Station's altitude

S Peracchi, S Guatelli, L T Tran and A B Rosenfeld
Centre for Medical Radiation Physics, University of Wollongong, NSW Australia

E-mail: sp009@uowmail.edu.au

Abstract. Latest researches predict the cosmic rays exposure during long space missions outside the Earth's geomagnetic sphere, as to the Moon and Mars, can double the cancer risk. The minimization of risk associated with radiation exposure in the Low Earth Orbit (LEO) is of great interest for astronauts of the International Space Station (ISS). This risk is due to radiation hazard and can be predicted by measurement of the dose equivalent produced by the radiation environment inside of the spacecraft. Since more than 15 years ago, the Centre For Medical Radiation Physics (CMRP), University of Wollongong, is developing silicon based microdosimeters as alternative solution to Tissue Equivalence Proportional Counters, for dose equivalent determination in the context of astronauts radiation protection. This work describes a Geant4-based study characterising the radiation environment encountered by astronauts inside the ISS. In particular, the energy spectra of different components of the cosmic radiation field, as Galactic Cosmic Rays (GCR), trapped protons and electrons and Solar Proton Event (SPE), has been simulated outside and inside the Columbus module where astronauts live. The simulated radiation environment inside of the ISS will be used to model the response of the 3D Mushroom Silicon On Insulator (SOI) microdosimeter recently developed at CMRP.

1. Introduction

Latest researches predict the cosmic rays exposure during long space missions outside the Earth's geomagnetic sphere, as to the Moon and Mars, can double the cancer risk. The need to characterize cosmic radiation and its effects on astronauts' health motivated the development of new sensible instruments capable to evaluate the dose at the cellular level, strongly damaged by radiation. The Centre for Medical Radiation Physics (CMRP) developed several generations of Silicon On Insulator (SOI) microdosimeters and the response of the first generation of microdosimeter was tested before when exposed to solar protons within a spacecraft [1]. In this study, the cosmic radiation field inside the International Space Station (ISS) has been characterized in detail. The calculated radiation field will then be used to fully characterize the response of the detector for personal microdosimetry in cosmic radiation, in the context of astronauts' radiation protection.

The Earth is surrounded by a harsh radiation environment, composed of three main sources: Galactic Cosmic Rays (GCR), coming from sources outside the Solar System, Solar Energetic Particles (SEP), emitted by the Sun during flares and coronal mass ejections, and the trapped radiation belts. The fluxes of particles change with the 11-years cycle of the Sun and depend strongly on the trajectory and altitude.

The environment scenario subject of this study is the Earth Orbit (LEO), where the International Space Station (ISS) is currently orbiting. It consists of three main components which are Galactic Cosmic Rays (GCR), Solar Proton Events (SPEs) and trapped protons and electrons. GCRs are mainly

composed of 87% of protons, 12% of alpha particles and 1% of heavy ions, whose energy can be up to 100GeV/n. SPEs are emitted by the Sun during solar flares and coronal mass ejections (CMEs) where it included electrons, protons and heavier charged particles of GeV [3, 4] energy range. The geomagnetic field of the Earth partially shields our planet from GCR and SPE, but also traps charged particles creating the so-called Van Allen belts. Energies are usually up to tents of MeV for electrons and hundreds of MeV for protons. Particles fluxes are influenced by the Sun activity as well the presence of the Earth's magnetic field: the first one is ruled by a cycle of 11 years of maximum and minimum peaks, whereas the second is linked to the cut-off rigidity, a quantity depending on geographical coordinates [2].
GCRs, SPEs and trapped particles hit the ISS and can traverse the shielding walls reaching astronauts' habitat, eventually producing nuclear products, e.g. neutrons, protons, pions, heavy ions deriving from fragmentation.

In this study the radiation field has been characterised in the ISS Columbus module as it is the astronauts' workspace. The type of the orbit and the altitude of ISS determinate the radiation environment, so real-time orbit details have been collected to simulate particles spectra at the present time of this study.

2. Materials and Methods

In the Geant4 simulation the Columbus module has been modelled based on specifications of the real Columbus structure and materials [3]. It is a cylinder with walls consisting of layers of different materials (multilayer). The wall thickness is approximately 7.9 cm. Main materials are Aluminium, Kevlar and Nextel. The material of the outer space and of the inner habitable space of the Columbus is vacuum.

The particles of the radiation field, incident on the Columbus module, have been modelled as generated from a random position on a sphere S of radius 1km. The Columbus module is at the center of S. The direction of the particles follows a cosine distribution. Such configuration has been demonstrated to model the isotropic radiation field typical of open space [4].In order to speed up the simulation times, the particles are generated towards the center of the sphere within a cone investing the Columbus.

The particles fluxes and energy spectra outside the ISS have been derived from SPENVIS, an online tool to model space environment and its effects developed by a consortium led by the Royal Belgian Institute for Space Aeronomy (BIRA-IASB) for ESA's Space Environments and Effects Section [5]. Considering a typical orbit of the ISS at the variable altitude between 401 and 408 km (perigee and apogee, respectively), with an inclination of 51°, we selected the following models in the solar minimum condition which corresponds to the maximum particles flux intensity:

- The NASA AP8/AE8 models for trapped protons and electrons in a range of energy up to 400 MeV and 7 MeV, respectively
- The CREME96 model for GCRs up to 100GeV/n
- The JPL-91 model for SPE protons with energy up to 500MeV

The differential fluxes of each radiation component have been used as input to the Geant4 simulation. The Geant4 *GeneratorParticleSource* has been used to model the radiation field investing the Columbus.

Both electromagnetic and hadronic physics interactions were modelled in the simulation with the following Geant4 Physics list *G4EmStandardPhysics_option3* and the *G4HadronPhysicsQGSP_BIC_HP*, respectively.

The output of the simulation is the fluxes and energy spectra of the particles emerging from the multilayer, inside the Columbus.

IOP Conf. Series: Journal of Physics: Conf. Series **1154** (2019) 012020 doi:10.1088/1742-6596/1154/1/012020

3. Results

Figure 1-5 show how particles spectra change behind the multilayer, when considering incident trapped protons and electrons, GCR protons and alpha particles and SPE protons, respectively.

In Figure 1 trapped protons are shown to survive mostly at energy above 10 MeV, accompanied by a significant production of secondary low energy neutrons. Because of their low kinetic energy (up to 7 MeV), trapped electrons get stopped inside the multilayer (Figure 2). Incident GCR protons with energy up to 100 GeV are attenuated in the lower energy range (they are fully absorbed up to 5 MeV). Indeed, Figure 3 shows a significant flux of neutrons from spallation reactions and a not negligible flux of alpha particles due to disintegrations of the light and heavy nuclei of walls materials. Similarly, incident GCR alpha particles with 2.6 MeV are completely shielded. Indeed, they produce a significant flux of secondary neutrons and protons of hundreds MeV (Figure 4). Finally, SPEs protons interacted with wall's material and produced a high fluence of secondary neutrons as shown in Figure 5.

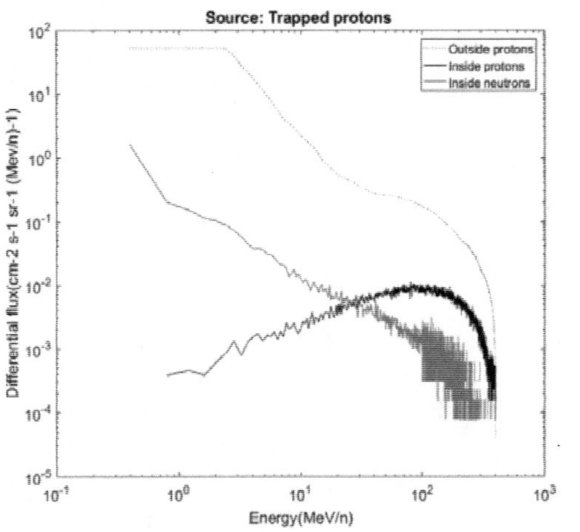

Figure 1. Kinetic energy spectra of incident trapped protons outside the ISS (grey), incident and secondary protons inside the ISS (black) and secondary neutrons entering the ISS (dashed line).

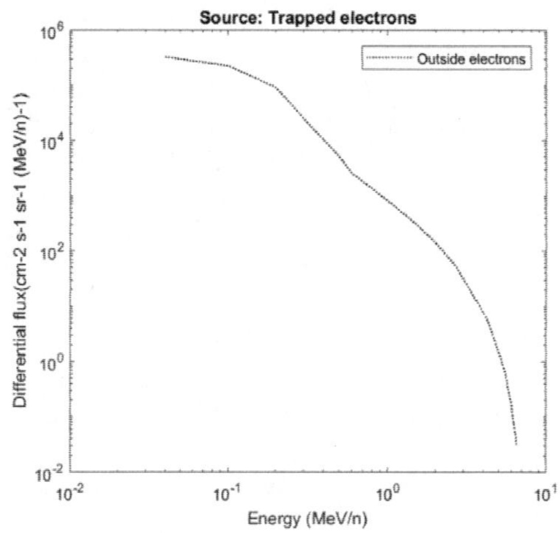

Figure 2. Kinetic energy spectra of incident trapped electrons outside the ISS. All of these have been stopped by the ISS's wall.

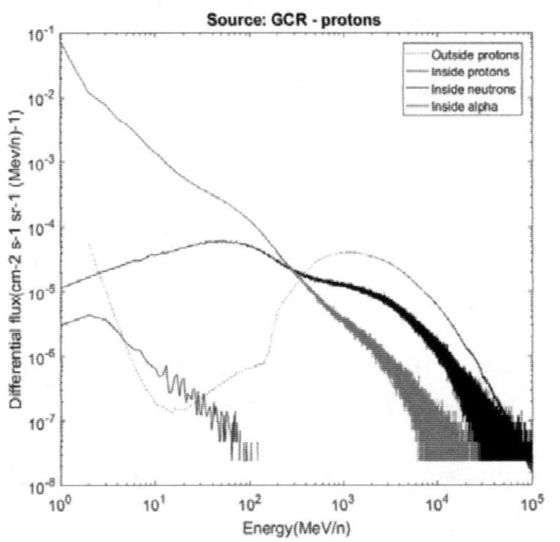

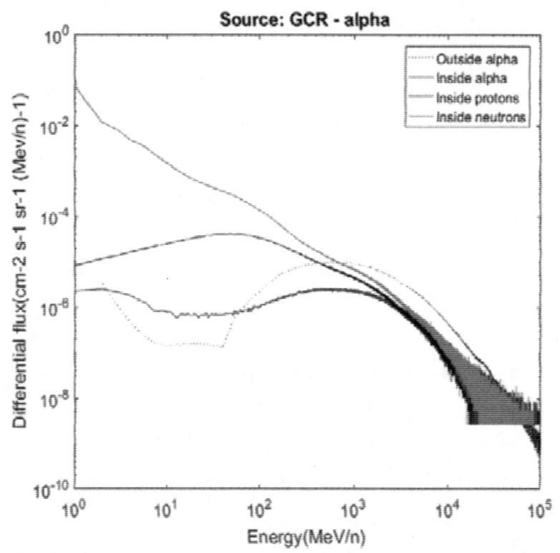

Figure 3. Kinetic energy spectra of incident GCR protons outside the ISS (grey), incident and secondary

Figure 4. Kinetic energy spectra of incident GCR alpha outside the ISS (grey), incident and secondary

IOP Conf. Series: Journal of Physics: Conf. Series **1154** (2019) 012020 doi:10.1088/1742-6596/1154/1/012020

protons inside the ISS (black) and secondary neutrons (big dashed line) and alpha particles (small dashed line) entering the ISS.

alpha particles inside the ISS (small dashed line) and secondary neutrons (big dashed line) and protons (black) entering the ISS.

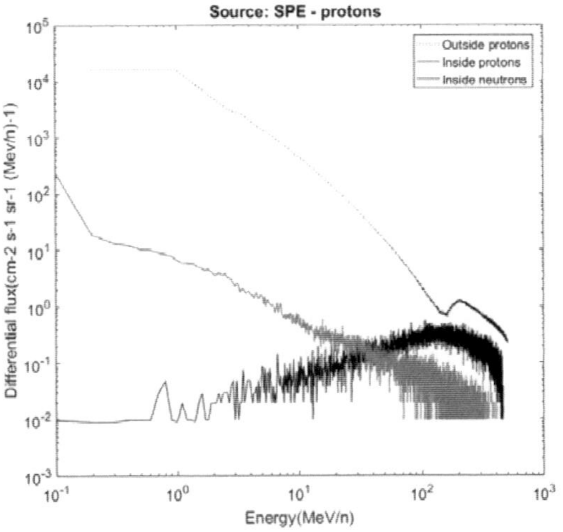

Figure 5. Kinetic energy spectra of incident SPEs protons outside the ISS (grey), incident and secondary protons inside the ISS (black) and secondary neutrons (big dashed line) entering the ISS.

4. Conclusions

The characterization of the LEO radiation environment where the ISS is orbiting and hosting several astronauts has been performed with the simulation of particles spectra coming from main radiation sources as trapped protons and electrons, GCR, and SPEs.

Particle spectra outside the Columbus module have been derived from SPENVIS showing high energy fluxes of protons and alpha particle, up to 100GeV/n, which represent a high risk for astronauts health. By means of Geant4, we simulated energy spectra inside the spacecraft behind a multi-layer shielding wall. The results shos that only the low energy component of the incident radiation is stopped in the ISS's wall, but high energy particles survive during the propagation through the ISS's wall, reaching the astronauts' habitat. Significant fluxes of secondary neutrons are produced, e.g. by spallation from protons. GCRs and SPEs seem to be the most dangerous source of radiation because of high energy particles.

In the next future simulations will be done to model the response of the 3D Mushroom microdosimeter, set inside the Columbus module, at different distances from the wall, using as input the mixed radiation field inside the ISS, calculated in this work. Based on the microdosimeter response, the dose equivalent for astronauts will be calculated.

5. References

[1] Wroe A J *et al* 2005 *IEEE T Nucl Sci* **52** 2591-6
[2] Smart D F and Shea M A 2008 Proceedings of 30th International Cosmic Ray Conference, Mexico **1** 737-40
[3] Destefanis R *et al* 2014 Universal Journal of Aeronautical & Aerospace Sciences **2** 80-113
[4] Santin G 2007 *Normalisation modelling sources* in *Geant4 tutorial* ESA/ESTEC
[5] https://www.spenvis.oma.be

IOP Conf. Series: Journal of Physics: Conf. Series **1154** (2019) 012021 doi:10.1088/1742-6596/1154/1/012021

4D robust optimization in pencil beam scanning proton therapy for hepatocellular carcinoma

T Pfeiler[1,2], D Ahmad Khalil[1,3], C Bäumer[1], O Blanck[4], M Chan[4], E Engwall[5], D Geismar[1,3], S Peters[1,3], S Plaude[1], B Spaan[2], J Wulff[1] and B Timmermann[1,3]
[1]West German Proton Therapy Centre Essen, Essen, Germany
[2]TU Dortmund University, Experimental Physics 5, Dortmund, Germany
[3]University Hospital Essen, Clinic for Particle Therapy, Essen, Germany
[4]University Clinic Schleswig-Holstein, Department of Radiation Oncology, Kiel, Germany
[5]RaySearch Laboratories AB, Stockholm, Sweden

E-mail: tina.pfeiler@uk.essen.de

Abstract. The treatment of moving targets is a challenging task using high conformal radiation techniques such as pencil beam scanning (PBS) proton therapy and requires adequate motion mitigation. Recent guidelines propose 4D robust optimization to mitigate motion artefacts in PBS therapy of thoracic malignancies. However, the availability of dosimetric analyses supporting this recommendation is limited and even non-existing for other tumour sites. The objective of this study was therefore to analyse the effectiveness of 4D robust optimization for hepatocellular carcinoma (HCC), representative for moving abdominal targets. These are usually less affected by uncertainties due to tissue heterogeneities than thoracic targets. 4D robustly optimized plans were compared with beam-specific margin plans for 6 HCC patients based on 4D dynamic accumulated doses (4DDD). 4DDD computations were conducted in RayStation with an experimentally validated routine including a site-specific beam time model. Contrary to expectations based on thoracic studies, 4D robust optimization did not yield a more homogeneous target coverage except for shallow targets close to the ribs. A clear advantage of 4D robust optimization is the sparing of normal tissue. The average dose to the normal liver could be reduced by up to 12%.

1. Introduction

Pencil beam scanning proton therapy (PBS) enables high-precision treatment for malignant tumours due to its steep dose gradients in beam direction. The utilization of the distinctive dose profile of protons in an intensity modulated approach (IMPT) allows sparing of organs at risk (OARs) even for complex anatomies. However, it is exactly this concept that makes PBS particularly susceptible to range uncertainties and organ motion. The interplay of organ motion and beam scanning may cause distortions of the intended dose distribution [1]. In order to exploit the full potential of PBS also for moving targets, motion mitigation techniques like robust optimization and modelling of interplay effects become crucial. Guidelines recently published for thoracic cancer [2] recommend 4D robust optimization to mitigate interplay effects in (intensity-modulated) PBS. However, only few exploratory studies exist [3-5] and dosimetric investigations of 4D robust optimization for different optimization algorithms and tumour sites are lacking, not least because an appropriate, site-specific beam time model is required for 4D dynamic accumulated dose (4DDD) computation. 4DDD computation is a prerequisite to evaluate whether the applied motion mitigation technique is adequate. Such a 4DDD tool is currently not available within a commercial treatment planning system (TPS).

This study investigates the potential of 4D robust optimization to reduce interplay effects and spare OARs in comparison to the conventional margin concept for hepatocellular carcinoma (HCC). HCC, which is the second leading cause of cancer related mortality worldwide [6], is selected as a model system for moving targets, because it is less affected by tissue heterogeneities and related range uncertainties than thoracic targets. 4DDD computations were performed with a customized and experimentally validated interplay effect routine in a clinical treatment planning environment [7].

2. Materials and Methods

2.1. Patient data and treatment planning

4DCT datasets from 6 HCC patients enrolled in a previous study [8] were included in the current work. The selective uptake of lipiodol by tumour cells after trans-arterial chemo-embolization was exploited as natural contrast and abdominal compression limited the tumour motion. The maximum motion vector length, defined as the difference between the dimensions of the clinical target volume (CTV) and the internal target volume (ITV), varied between 0.7 cm and 1.3 cm. The tumour size ranged from 3 cm^3 to about 500 cm^3 (average 160 cm^3).

Two-field PBS treatment plans were generated in a research version of RayStation 6 (RaySearch Laboratories AB, Stockholm, Sweden) using the pencil beam dose algorithm and 4D robust optimization based on worst case scenarios (4D plans). A uniform dose of 63 Gy_{RBE} (D_{pres}) was prescribed to the CTV in 15 fractions according to Ref. [9] allowing for intensity modulations within single fields. The robust optimization settings were chosen to achieve plan robustness against 2 mm setup error, 5% range uncertainty and morphological changes in 10 respiratory phases. For comparison purposes, corresponding non-robustly optimized PBS plans were created on the end exhalation phase based on beam-specific planning target volumes using a single field uniform dose approach (SFUD plans). The uncertainty parameters above were used as margins in this case, whereby the internal margin (accounting for target motion) was derived from 10 respiratory phases. For both, 4D and SFUD plans, the energy layer spacing and the spot distance were set automatically with a scale factor of 1.1 and 0.8, respectively. The energies ranged from 100 MeV to 195 MeV, whereby range shifters were applied, when necessary. The air gap between a range shifter and the patient varied from 2 cm to 5.4 cm.

2.2. Interplay effect evaluation

An interplay effect routine tailored to 4DDD computations of irradiations with an IBA Proteus®Plus proton therapy system (IBA, Louvain-La-Neuve, Belgium) in RayStation forms the basis for all plan analyses as described in a former project [7]. The time structure of the field delivery is provided by an empirical beam time model and used to determine the delivery start time of a beamlet (spot). Based on the specified respiratory cycle length and starting phase of the irradiation, the application of each spot can be assigned to one of the respiratory phases of the 4DCT. For every phase *i*, a fraction dose $d_{fx,CTi}$ is calculated and subsequently mapped to the reference phase *j* using the ANAtomically CONstrained Deformation Algorithm, ANACONDA [10]. The summation over all deformed doses $d_{fx,CTi}^{Ref}$ and the dose delivered within the reference phase $d_{fx,CTj}^{Ref}$ yields the fractional 'interplay dose distribution' $d_{fx} = d_{fx,CT1}^{Ref} + d_{fx,CT2}^{Ref} + \ldots + d_{fx,CTr}^{Ref}$, where *fx* indicates the fraction number and *r* the total number of respiratory phases. The total delivered dose *D* is given by the sum over all fraction doses: $D = \sum_{fx=1}^{N} d_{fx}$.

The outcome of a single fraction was analysed based on 60 simulations per plan with varying starting phases and energy layer switching times in accordance with Ref. [7]. A whole treatment course, consisting of 15 fractions, was modelled with a reduced set of 5 simulations per fraction (75 simulations per plan). The homogeneity index $HI = (D_5 - D_{95})/D_{pres}$ and the percentage over- and underdosage $V_{107,95} = V_{107} + (100 - V_{95})$ served as evaluation criteria due to their sensitivity to interplay effects. For all evaluations, a constant respiratory period of 4 s was chosen.

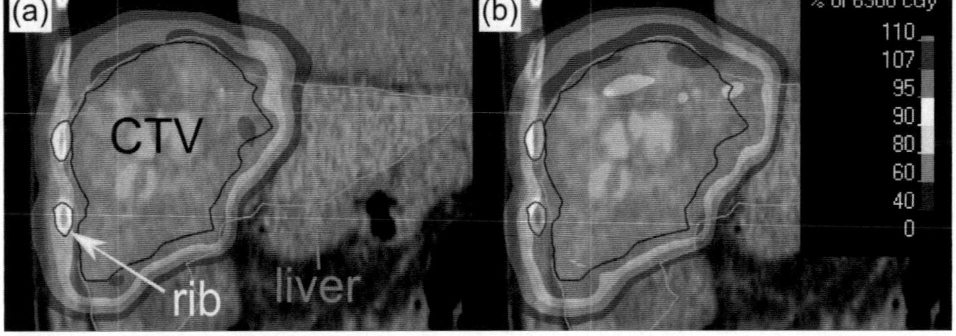

Figure 1. Single fraction 4D dynamic accumulated dose for a 4D robustly optimized plan (a) and a non-robustly optimized plan (b) (coronal view).

3. Results

Figure 1 shows exemplarily the coronal view of a 4DDD distribution of a single fraction for a 4D and a SFUD plan. For a single fraction, the HI of the CTV was on average 9.2% for 4D and 8.8% for SFUD plans. The standard deviation (SD) varied from 0.6% to 2.1% (4D) and 0.8% to 1.7% (SFUD), respectively (see Fig. 2). 4D plans exhibited on average a $V_{107,95}$ value of 5.9% within the CTV and SFUD plans 5.0%. The corresponding SD ranged from 1.2% to 5.6% (4D) and 2.2% to 4.1% (SFUD), respectively. No correlation between the size of the CTV or the motion amplitude and the severity of interplay effects was found. Slightly smaller HI and $V_{107,95}$ values were observed for 4D plans when the CTV was located close to the ribs whereas SFUD plans yielded better 4DDD results for centrally-located targets in the liver (see Fig. 2).

OARs close to the target, such as ribs and the healthy liver tissue, could be better spared when using 4D robust optimization (see Fig. 1). For shallow targets, the rib volume receiving more than 60 Gy_{RBE} was on average 8 times larger for SFUD plans than for 4D plans leading to a probability of rib fracture of up to 6.3% according to Ref. [11]. The average dose to the healthy liver tissue could be decreased by 0.6% to 12.0% (average 6.9%) with 4D robust optimization.

Simulating the whole treatment course over all fractions, the HI decreased to less than 5% for both treatment techniques as can be seen in Fig. 2. The average HI of 3.2% (4D) and 3.1% (SFUD) was only a little higher than for the static case (2.0% and 2.1%). $V_{107,95}$ approached 0% in all cases.

4. Discussion

On average, the interplay effect was slightly less pronounced for SFUD plans than for 4D plans. The reason is presumably the larger volume covered by 95% of D_{pres} in SFUD plans than in 4D plans (average 18.8%). This potentially reduces the over- and underdosage at the edge of the CTV. Further, SFUD plans exhibit a uniform dose per field and inhomogeneities in one of the fields might be better compensated which is expected to improve robustness against motion. However, for shallow targets (1, 3 and 5 in Fig. 2), 4D plans led to slightly more homogeneous target coverage. Here, the 4D optimization algorithm avoided irradiation through the ribs and CTV regions directly behind the rib were covered by the other field.

The study demonstrated that 15 fractions were sufficient to mitigate interplay effects within the CTV for both 4D and SFUD plans. The dose to normal tissue could be significantly reduced in 4D plans since robust optimization adjusts the irradiation volume to density changes unlike the more conservative margin concept. The advantages (better OAR sparing and slightly less interplay effects for shallow targets) and disadvantages (slightly stronger interplay effects for centrally-located targets and long computation times) of 4D robust optimization must be weighed up on a case by case basis.

Treatment planning options that were not specifically addressed in this study - but have the potential to further mitigate interplay effects within a single fraction - are a larger number of treatment fields, rescanning and the increase of the air gap between range shifter and patient to enlarge the spot size [12]. This would be at the expense of treatment time and a sharp lateral dose fall-off.

All 4DDD computations rely on the respiratory motion information derived from planning 4DCTs and might lose their validity if motion trajectories changes occur during treatment. As a consequence, great care should be taken when transferring knowledge from research to clinic where individual 4DDD evaluations based on a planning 4DCT might lead to a false sense of security.

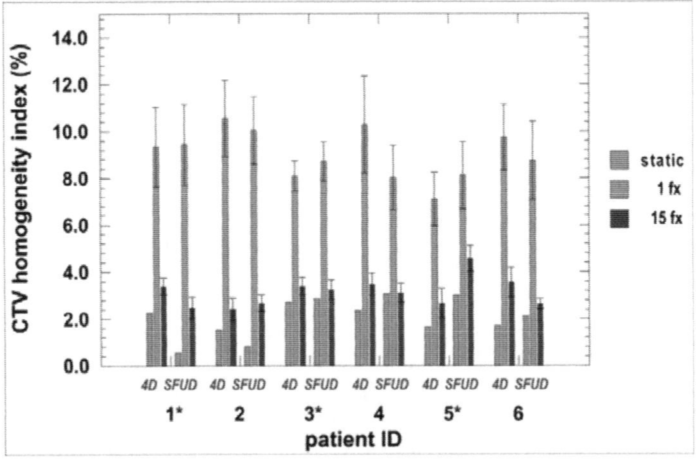

Figure 2. Comparison of the CTV homogeneity index for 4D robustly optimized (4D) and non-robustly optimized (SFUD) plans: 4D dynamic accumulated dose distributions of 1 and 15 fractions (fx) and the static dose were evaluated. Error bars indicate the standard deviation for 60 simulations (1 fx) and 5x15 simulations (15 fx). The CTV is more centrally-located in the liver for patients 2, 4 and 6 and shallow for 1, 3 and 5 (marked by an asterisk).

5. Conclusion

Using the specified uncertainty criteria, 4D robust optimization was not found to be superior to the conventional margin concept in terms of reducing interplay effects for the HCC patients under consideration. In the case of centrally-located targets, HI and $V_{107,95}$ were even slightly better for non-robustly optimized plans considering a single fraction. However, in the case of shallow targets close to the ribs, 4D robust plans were shown to be slightly beneficial in terms of target coverage. Moreover, robust optimization reduced the 4DDD to close-by OARs which can be of importance for critical cases. The study showed that dose inhomogeneities were averaged out for the applied fractionation scheme of 15 fractions for both treatment planning techniques. The results could differ for patients treated under other conditions, e.g. without abdominal compression or lipiodol contrast. This needs to be further investigated in a larger patient cohort.

6. References

[1] Bert C *et al* 2011 *Phys. Med. Biol* **56** R113-44
[2] Chang J Y *et al* 2017 *Int. J. Radiat. Oncol. Biol. Phys.* **99** 41-50
[3] Liu W *et al* 2016 *Int. J. Radiat. Oncol. Biol. Phys.* **95** 523-33
[4] Newpower M *et al* 2016 *Med. Phys.* **43** 3504-5
[5] Yu J *et al* 2016 *Med. Phys.* **43** 1111-8
[6] Ferlay J *et al* 2015 *Int. J. Cancer* **136** E359-86
[7] Pfeiler T *et al* 2017 *Z. Med. Phys.* **28** 121-33
[8] Chan M K *et al* 2016 *Strahlenther. Onkol.* **192** 92-101
[9] Bush D A *et al* 2011 *Cancer* **117** 3053-9
[10] Weistrand O *et al* 2015 *Med. Phys.* **42** 40-53
[11] Kanemoto A *et al* 2013 *Acta. Oncol.* **52** 538-44
[12] Grassberger C *et al* 2013 *Int. J. Radiation. Oncol. Biol. Phys.* **86** 380-6

IOP Conf. Series: Journal of Physics: Conf. Series **1154** (2019) 012022 doi:10.1088/1742-6596/1154/1/012022

A comprehensive phantom with multi-detector inserts for pre-treatment quality assurance in stereotactic ablative radiotherapy

P Ramachandran[1,2,3] N Esen[1,2], A Tajaldeen[1,2], M Geso[2], D Taylor[1], P Archer[1] and T Kron[1,2,3]
[1]Peter MacCallum Cancer Centre, Victoria, Australia
[2]RMIT University, Victoria Australia
[3]Monash University, Victoria Australia

E-mail: Ramachandran.Prabhakar@petermac.org

Abstract. Stereotactic ablative radiotherapy (SABR) is an advanced high precision radiotherapy procedure where a high dose of radiation is delivered in a few fractions. Pre-treatment quality assurance (QA) is conducted prior to treatment to ensure the planned dose is delivered precisely. Several types of detectors have been explored for the use of pre-treatment QA in SABR. This study discusses the design of an in-house phantom with inserts designed to accommodate multiple detectors such as ion chambers, films, alanine, optically stimulated luminescence (OSLs), thermoluminescence dosimeter (TLDs), diodes, diamond detectors, and presage dosimeter. The study also presents some of the measurement results performed for selective treatment sites. The inserts were designed using Perspex and 3D printed materials. The Presage dosimeter provided a three-dimensional view of the delivered dose for SABR plans and matched closer to the planned dose. Most of the detectors/dosimeters explored in this study demonstrated measured results within 3% of the planned dose.

1. Introduction

Stereotactic Ablative Radiotherapy (SABR) also known as Stereotactic Body Radiotherapy (SBRT) is an advanced form of radiotherapy technique employed to deliver high doses of radiation in a small number of fractions, typically between one and five. Some of the salient features of SABR are a) less dose spillage b) high dose to ablate a tumour c) high precision and accurate dose delivery d) highly conformal e) aimed to improve overall survival and f) reduced overall treatment time. In the current era, SABR is extensively used in the treatment of the following sites: Lung, Liver, Spine, Kidney, Bone metastasis and Prostate. Figure 1 highlights the requirement of SBRT/SABR wherein precision (similar to stereotactic radiosurgery (SRS)) and image guidance are the key pillars of SABR technique. High dose of radiation delivered in a short period requires stringent quality assurance (QA) of the image guidance system and patient positioning systems. In addition to this, pre-treatment patient-specific quality assurance needs to be conducted prior to treatment delivery. Several guidelines have been published emphasising the importance of quality control and mandatory QA procedures conducted as part of the SABR treatment process [1-4]. Pre-treatment patient-specific quality assurance involves both absolute and relative dosimetry analogous to the QA conducted for intensity modulated radiation therapy (IMRT)/volumetric modulated radiation therapy (VMAT) [5, 6]. Several dosimeters have been proposed for patient-specific QA, among all, ionisation chamber and film are considered to be the gold standard. Measurements could be classified as a) 1D or point dosimetry, b) 2D employing films, array detectors (ion chambers, diodes etc.,), c) 3 dimensional, using gel/presage dosimeters [7-9] d) 4D use of 3D dimensional detectors simulating tumour movement. Though ionisation chambers are the gold standard, of late several other detectors including thermoluminescence detectors (TLD), alanine, optically stimulated luminescence dosimeters (OSL), diodes, diamond detectors, scintillation detectors etc., have been explored in radiotherapy.

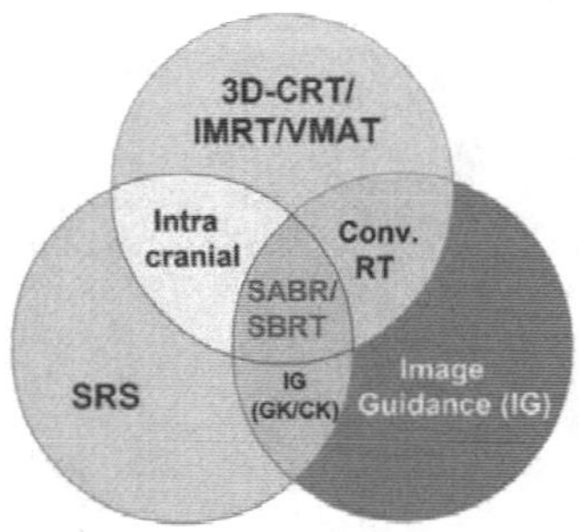

Figure 1. SABR and other related radiotherapy technologies

In this study, we present an in-house phantom originally designed to house an ionisation chamber and films, which has been customised to accommodate multiple detectors such as TLD, alanine, OSL, diamond detectors, and presage 3D dosimeters for patient-specific QA of SABR/SBRT patients.

2. Materials and Methods

An in-house Perspex phantom known as Rod phantom was originally designed to accommodate a CC13 ionisation chamber. The Rod phantom mimics the shape of the thorax and has options to place films and TLDs in the transverse plane. The CC13 chamber enables the measurement of absolute dose. The relative dosimetry is obtained by using a film placed in the transverse plane. Figure 2a demonstrates an outline of the rod phantom and 2b shows the top view of the phantom with multiple slabs placed aside. Figure 2c shows a slab with embedded pins (highlighted in red ellipses) used to identify the orientation of the films and 2d exhibits the chamber entry port and the location of the film slab. Prior to performing patient-specific QA, chamber calibration is performed by exposing a known dose to the ionisation chamber at patient specific beam energy. Figure 3 illustrates the steps involved in patient-specific QA which involve exporting a SABR plan onto the rod phantom and then recalculating the dose. The dose measured on the linac is compared to the calculated dose.

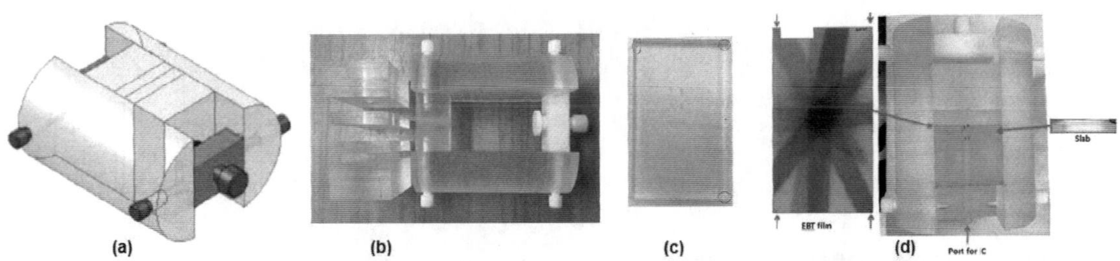

Figure 2. (a) Rod phantom, (b) phantom shell top view and slabs, (c) TLD slab with pins, (d) assembled phantom top view.

At our centre, most of the SABR QAs are performed using the rod phantom which enables the physicists to acquire the absolute and relative dosimetry in one measurement. Multiple slabs in the phantom enabled us to explore other commercially available detectors. To accommodate these dosimeters inside the phantom, a Perspex slab with a circular hole as shown in Figure 4 b was designed to accommodate different tissue equivalent 3D printed inserts, Perspex concentric rings and Perspex spacers. The 3D inserts were designed in ViaCAD^{TM} 2D/3D software, and output files were exported in .stl format. The inserts were made of Acrylonitrile Butadiene Styrene (ABS) and printed using a CreateBot™ 3D printer.

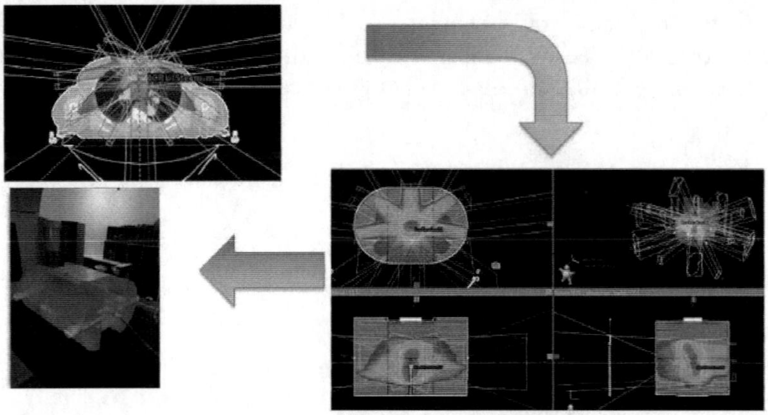

Figure 3. Workflow of patient specific QA.

Alanine was one such dosimeter that has near tissue equivalence with very low energy dependence and possesses excellent isotropic response. Alanines used in this study have dimensions of 4.80±0.3mm by 3.00±0.3mm with 96% of alanine and 4% binder and are made at the Synergy Health (SH) and National Physical Laboratory (NPL). These dosimeters can be used between 10 Gy and 150 kGy with linear dose response up to 3kGy and are highly suitable for hypo-fractionated radiotherapy.

A second insert was printed to accommodate TLD 100H chips. The TLD 100H ((LiF: Mg, Cu, P) chips have an effective atomic number of 8.2 with a good linear response up to 10 Gy, less energy dependence as compared to TLD 100. The TLD 100H chips used in this study were of 3.2 mm x 3.2 mm x 0.89 mm in size.

Nanodots optically stimulated luminescence (OSL) are widely used in personnel dosimetry are gaining popularity in radiotherapy. The nanodot OSLs work on the same principle as that of the TLDs, except the nanodots have to be read with the aid of an optical light source/laser beam instead of heat. The Nanodot OSLs consists of a 4 mm diameter dot of aluminium oxide doped with Al_2O_3:C, housed inside a square package. A 3D ABS insert was printed on the 3D printer to accommodate the nanodot OSLs and the insert was placed inside the Perspex slab and stacked inside the rod phantom. The exposed nanodot OSLs were read in Microstar™ reader (Landauer®).

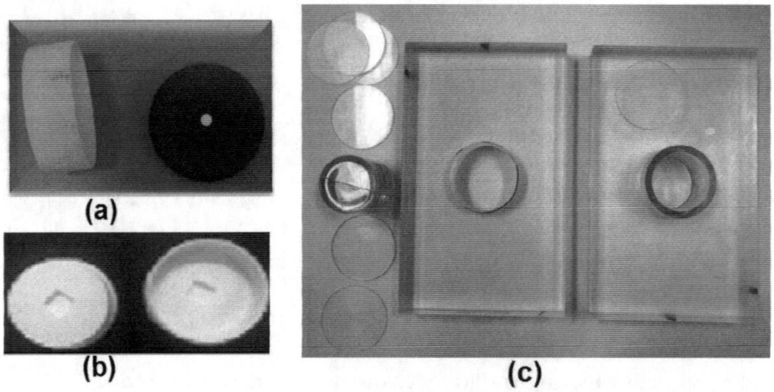

Figure 4. (a) 3D printed inserts for alanine, (b) for OSL, (c) perspex slabs with circular holes and spacers.

The PTW microDiamond type 60019 is a small volume detector of 0.004 mm^3 well suited for small field dosimetry with near tissue-equivalence property and negligible dose rate dependence. An insert in Perspex was also designed for housing the Diamond detector, located in the same position as the CC13 ionization chamber.

The PRESAGE is a solid radiochromic 3D dosimeter widely used across a range of clinical applications [10-12]. It is composed of a transparent polyurethane matrix, Triarylmathane leuco dye and halogenated carbon radical initiators. It has several advantages over point dosimeters and even the commercially available 2D array detectors. The dose can be recorded in 3D space at a high resolution, and it is a lot easier to handle when it comes to measurement and insensitive to room light. Upon exposure to radiation, it undergoes a colour change which is highly dependent on the amount of exposed radiation. The presage dosimeters were prepared and poured into the cylindrical plastic container. Multiple Perspex slabs with circular holes were used to accommodate the Presage dosimeters. In this study, a maximum of 8 cm diameter Presage dosimeters were tested for SABR QA. Small Presage dosimeters could also be accommodated in this phantom by adding additional circular Perspex ring and spacers together. The images were processed in ImageJ software, and the dose maps were generated in PolyGeVero software.

SABR plans were generated in Eclipse™ treatment planning system (Varian Medical Systems) and verification plans were generated on the Rod phantom and transferred to linac for measurement. The phantom was positioned to the centre of target volume and measurement was made for all of the above mentioned detectors/dosimeters.

3. Results

Figure 5 shows the ratio of the ion chamber measurement results to planned dose. Non-SABR lung cases include spine, sternum and metastatic sites such as scapula, hip and ribs. Due to high dose gradients in complex SABR cases, some of the fields exceeds 3% but were within 5% of the field doses. The use of EBT3 films in rod phantom and the results have been demonstrated in our previous study [13] The results of alanine, TLD and nanodot OSLs showed very good agreement (<3%) with the Eclipse planning system.

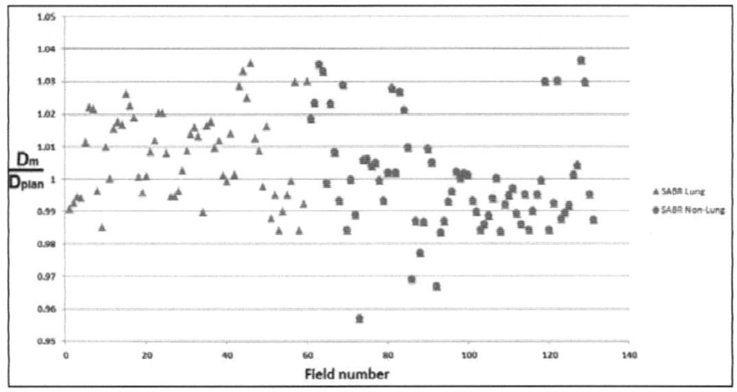

Figure 5. Ion chamber results of 131 fields for SABR and non-SABR lung patients.

Figure 6 demonstrates the diamond results for a SABR spine patient proving the efficacy of the use of small sensitive volume for SABR QA. Presage dosimeter proved to be a valuable dosimeter for SABR QA to visualise the 3D dose distribution and moreover it is well suited for small fields as most of the optical CT scanners have some limitations with the size of Presage dosimeter used to acquire Optical CT scans.

Beam	Measured (cGy)	TPS (cGy)	Diff. (%)
Ant 0	315.83	313.80	-0.65
OBL 95	226.00	221.90	-1.85
OBL 265	261.64	263.80	0.82
OBL 150	107.44	107.00	-0.41
OBL 210	102.75	102.30	-0.44
OBL 60 C20	175.73	173.00	-1.58
OBL 50 C340	200.93	197.80	-1.58
OBL 310 C20	192.84	193.60	0.39
OBL 300 c340	177.43	178.00	0.32

Figure 6. Diamond measurement results compared to TPS.

4. Discussion
Pre-treatment SABR QA is an essential requirement in SABR treatment workflow that provides confidence in the accuracy of the planned dose before treatment delivery. Currently, there are several detectors/dosimeters available on the market. However, most of the detectors require specially designed phantom for measurements. In this study, we've exhibited a phantom that could accommodate most of the commercially available detectors and almost mimics the shape of the thorax. The phantom has been designed to reduce physics QA by performing both point dosimetry and 2D / 3D dosimetry simultaneously.

5. Conclusion
In this study, we demonstrated the design of a unique in-house phantom which could accommodate multiple detectors that include point dosimeters, 2D and 3D dosimeters. The phantom is quite robust, and our results demonstrate its use for performing both absolute and relative dosimetry for SABR patients.

6. References
[1] Benedict S H *et al* 2010 *Med. Phys.* **37** 4078-101
[2] Galvin J M e*t al* 2008 *Int. J. Radiat. Oncol. Biol. Phys.* **71** S122-5
[3] Solberg T D *et al* 2012 *Pract. Radiat. Oncol.* **2** 2-9
[4] Kron T *et al* 2018 *Br. J. Radiol.* 2017
[5] Greer P B *et al* 2007 *Med. Phys.* **34** 4389-98
[6] Vial P *et al* 2008 *Med. Phys.* **35** 4362-74
[7] Baldock C 2009 *J. Phys.: Conf. Ser.* **164** 012002
[8] Mather M L *et al* 2003 *Phys. Med. Biol.* **48** N269-75
[9] Healy B J *et al* 2003 *Med. Phys.* **30** 2282-91
[10] Brown S *et al* 2008 *Appl. Rad. Isotop.* **66** 1970-1974
[11] Gorjiara T *et al* 2011 *Med. Phys.* **38** 2265-74
[12] Vial P *et al* 2006 *Phys. Med. Biol.* **51** 5517-38
[13] Ramachandran P *et al* 2017 *J. Med. Phys.* **42** 251-257

Brachytherapy utilising miniaturised X-ray tubes – clinical implementation and future applications

P Ramachandran[1,2,3], D Wanigaratne[1], J Hagekrykou[1], K Roozen[1], M Geso[2], T Kron[1,2,3] and S David[1]
[1]Peter MacCallum Cancer Centre, Victoria, Australia
[2]RMIT University, Victoria Australia
[3]Monash University, Victoria Australia

E-mail: Ramachandran.Prabhakar@petermac.org

Abstract. Electronic brachytherapy is an advanced radiation treatment technique specifically designed to deliver high doses of radiation inside or very close to the tumor-bearing tissues. Unlike the traditional radionuclide-based brachytherapy, electronic brachytherapy uses a miniaturised x-ray tube that can produce radiation when energised. Most of the electronic brachytherapy systems operate at 50 kVp, therefore posing less radiation exposure to both patients and staff when compared to the standard Cobalt-60 or Iridium-192 based brachytherapy systems. Electronic brachytherapy systems have broader applications that include the treatment of skin, brain, breast, spinal metastasis, endometrium, and cervix. The Axxent Xoft brachytherapy system is a unique system which houses a miniature X-ray tube of 2.25 mm diameter at the tip of a flexible catheter. As a mobile unit, the Xoft system requires less shielding and can be operated in remote centres with minimal radiation shielding facilities. The dose fall-off characteristics mimic those of the low energy isotopes, yet the unit still maintains the high dose rate property of an Ir-192 source. This study details the implementation and commissioning aspects of the Axxent system for intra-operative radiotherapy in Breast Cancer and future applications. Electronic brachytherapy systems have the potential to replace conventional radionuclide-based brachytherapy sources and systems.

1. Introduction

Brachytherapy is one of the oldest forms of radiation therapy with more than 100 years of clinical applications [1-3]. It uses sealed sources placed inside or close to cancerous tissues and is commonly used in the treatment of Head and Neck, breast, cervix, prostate and skin cancers. Brachytherapy commenced with the use of low dose rate radioactive sources, but over the last few decades, high dose rate brachytherapy machines have become a standard system for the delivery of brachytherapy treatments. The International Commission of Radiation Protection (ICRP) published a report in 2005 that reviewed all the radiotherapy accidents associated with high dose rate brachytherapy [4]. The report showed that radionuclide-based brachytherapy resulted in 500 accidents including one death along the entire chain of procedures from source packing to dose delivery. Some of the accidents were related to loss of sources, radiation leakage and inaccurate calibration of sources. The concept of using miniaturised x-ray tubes (electronic brachytherapy) as an alternative source for brachytherapy procedures was proposed by Alan Sliski of Photoelectron Corporation in collaboration with MIT [5]. Since then, the design of miniaturised tubes has led to x-ray tubes mimicking closer to the size and shape of the sources used in high dose rate remote afterloading machines. The use of miniature X-ray

IOP Conf. Series: Journal of Physics: Conf. Series **1154** (2019) 012023 doi:10.1088/1742-6596/1154/1/012023

tubes has several advantages over the conventional radionuclide-based brachytherapy. These low kilovoltage x-ray sources require relatively less shielding compared to Co 60, Cs 137, Ir 192 based radionuclide brachytherapy. The main highlighting feature of X-ray based sources is the absence of radiation when the source is not in use. It is highly expected that the introduction of electronic brachytherapy (eBx) systems would eliminate most of the accidents associated with radionuclide brachytherapy since the electronic brachytherapy system requires a dedicated system to deliver radiation and there is no concern about radiation leakage in the off condition. Also, there is no radioactive waste associated with X-ray source-based brachytherapy systems and no concerns with source transportation; unlike radionuclides where it requires a special license for source transportation. To our knowledge, there are around six different types of electronic brachytherapy systems currently available in the market, which includes Intrabeam (Zeiss), Xoft (iCAD), Papillon, Photoelectric therapy, Esteya and SRT 100. Most of these systems operate between 50 and 100 kVp. Intrabeam and Xoft system are the most popular ones which have wider application to many treatment sites. The Xoft anode is made of a tungsten film on an aluminium nitride substrate with yttria binder which emits K-lines at 14.9, 16.7 keV and tungsten L lines at 8.4, 9.8 and 11.3 keV. The Zeiss anode is made of a gold film on an beryllium substrate and the gold lines occur at 9.7, 11.5 and 13.4 keV. When compared to intrabeam, the Xoft source almost mimics an Ir-192 source in terms of dose rate and it is the only system that can be used to treat cervix and vagina due to its peculiar design closer to a wired radionuclide source such as Ir-192 source. This system can be used to treat cancers of the skin, breast, vagina and cervix. The maximum anode current, at 50 kVp is 300 μA and the output can be measured with an in-build electrometer and well-type ionisation chamber. This paper discusses the implementation and commissioning aspects of the Axxent Xoft system for breast intra-operative radiotherapy.

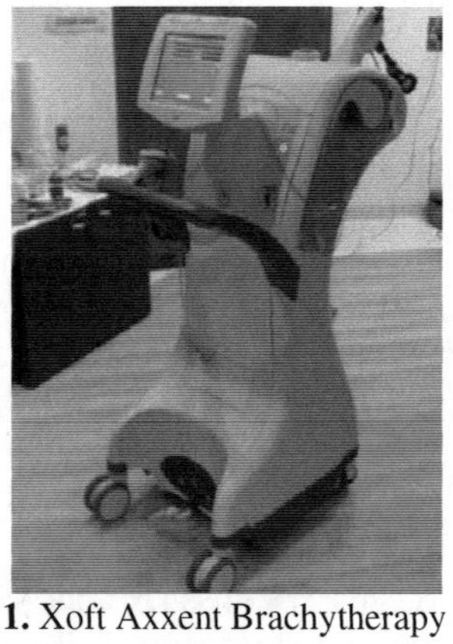

Figure 1. Xoft Axxent Brachytherapy System

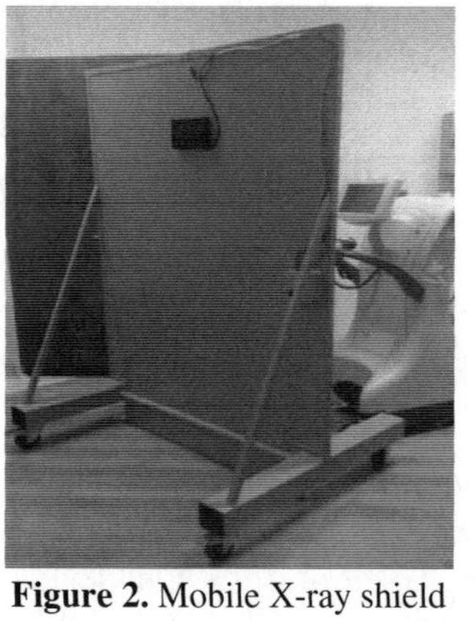

Figure 2. Mobile X-ray shield

2. Materials and Methods

In early 2016, Monash Cancer Centre, one of the campuses of Peter MacCallum Cancer Centre, received an Electronic Brachytherapy system (Figure 1) from iCAD Incorporation to evaluate the machine for Intra-operative breast radiotherapy (IORT) procedures. Prior to conducting any tests on the machine, the Department of Physical Sciences, Peter MacCallum Cancer Centre contacted Department of Health & Human Services, Victoria (DH&HS) and submitted an application with a detailed shielding calculation report for Radiation License to operate the machine. We built two Mobile Lead X-ray shields to reduce the dose to staff operating the machine inside the operating theatre (Figure 2). After obtaining the approval for one of our theatres at Moorabbin Hospital, physical sciences department evaluated the system for its suitability in Intra-operative procedures by conducting a series of experiments to measure the X-ray energy spectra, dose distribution, HVL and spatial accuracy of the source position during treatment. We measured the energy spectra of the 50 kVp X-ray source using an AMP-TEK XR-100T spectrometer. The AMP-TEK spectrometer is a CdTe

IOP Conf. Series: Journal of Physics: Conf. Series **1154** (2019) 012023 doi:10.1088/1742-6596/1154/1/012023

Gamma / X-ray detector with a PX4 digital pulse processor connected to the detector captures the signals from the spectrometer. To assess the maximum energy of the beam, we calibrated the spectrometer at low energies using γ-rays from radioisotopes, Tc-99m and Iodine-125. We measured the spectra for two different conditions (i) Energy spectra at 1 m for a bare X-ray source in air (ii) Energy spectra close to a 3-4 cm spherical balloon applicator. The spatial dose distribution of the x-ray source was measured using EBT3 films (Ashland Inc). The positional accuracy of the X-ray source was performed by exposing an EBT3 film underneath a perspex marker block. The lines engraved in the perspex block creates an impression on the EBT3 film. Using inhouse developed image processing software (FilmReader), the film was analysed for positional accuracy of the source at the marked positions. The output was measured using the inbuilt electrometer and well-type chamber. A cubical phantom was designed in a 3D printer and the dose was measured on the surface of the balloon applicator at specific positions and the measured dose was compared to manual calculation. Upon evaluating the system, Peter MacCallum Cancer Centre in collaboration with Moorabbin Hospital participated in an international Intraoperative radiotherapy breast cancer clinical trial mainly targeting early breast cancer patients. The prescription dose is 20 Gy in single fraction delivered to the surface of the lesion. This procedure involved a dedicated team which includes Surgeons, Radiation Oncologist, Physicists, Nursing Staff and Radiation Therapists. Figure 3 displays the workflow and roles of different disciplines in accomplishing this task. Prior to IORT procedure, physics performs day of treatment QA which includes checking coolant water level (should be greater than 1/3rd of volume), water clarity, coolant tubing assembly for restriction, aberration and kinks, x-ray source catheter tip for discolouration / deformation / cracks / brittleness, measure air kerma strength and compare with the previous measurement, calibration ratio, pullback arm interlock, wheel-lock interlock, applicator interlock, source connection interlock, positional and dwell time accuracy. Physics also conducts the pre-treatment QA which includes verification of patient details, recording initial CED/foley catheter volume and call out final applicator balloon volume, symmetricity of the fill volum around the applicator lumen, recording skin to balloon surface clearance and use of any rigid shields, selecting appropriate plan from library, measure source air kerma strength, measure applicator and source length, calibration ratio, check corrected dwell times, placement of flexishield, ensure pullback arm test is ok, ensure non-essential leaves the theatre and verify the stated volume matches with the post treatment volume. A total of 30 breast IORT patients have been treated to date using the Xoft Axxent electronic brachytherapy system.

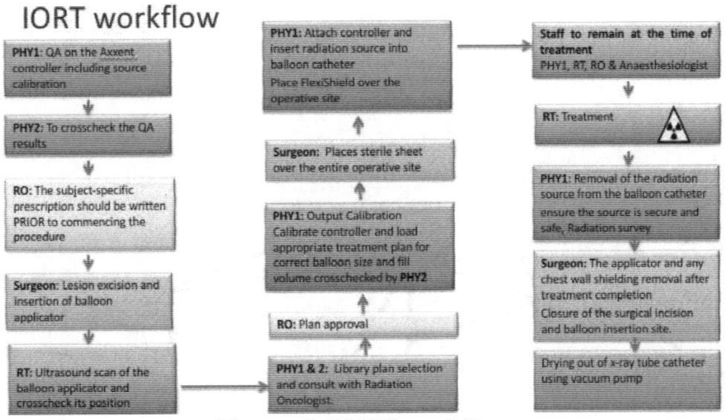

Figure 3. IORT workflow.

3. Results

Figure 4 shows the EBT3 film exposed for a pre-programmed single source position and isodose curves generated using the Film reader software. The film was exposed to a single dwell position in a spherical applicator in a custom built phantom. In figure 4 an image of the raw film is shown on the left that darkens from yellow to black as it absorbs radiation. It was assumed that the spectrum from a superficial unit was similar to that of the Axxent Xoft, so the film was scanned and calibrated using the dose distribution from this. A series of small EBT films from the same batch was exposed to a known dose of superficial radiation and the average scanned pixel value recorded as a mapping. This mapping was then applied to the original film to get an estimation of the dose for the Xoft Axxent. The resulting isodose plot shows a steep dose falloff dropping from about 35 Gy down to 0.2 Gy in a distance of just 3 cm from the source. This indicates that the Xoft Axxent system is very good at sparing surrounding healthy tissue outside the target volume.

Figure 5 shows the processed film showing preprogrammed source positions in steps of 0, 5, 15 and 25 mm and Table 1 shows the differences between the pre-programmed and measured positions. The energy spectra measured using Amp-TEK spectrometer indicated the maximum energy of the spectrum at 50 kVp [6]. The air kerma strength measured prior to treatment for 30 IORT patients was within 2%.

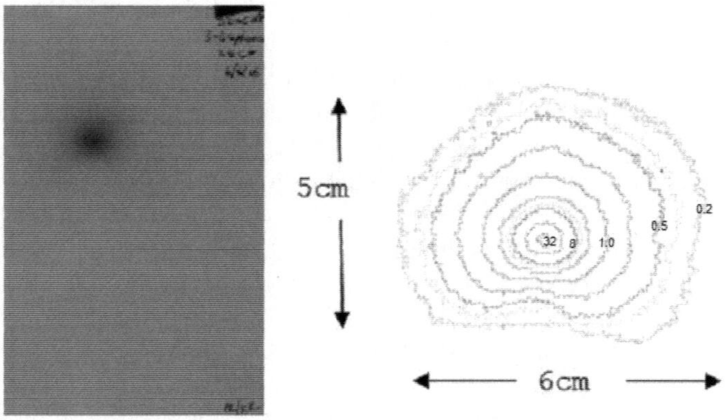

Figure 4. Exposed EBT3 film and isodose curves.

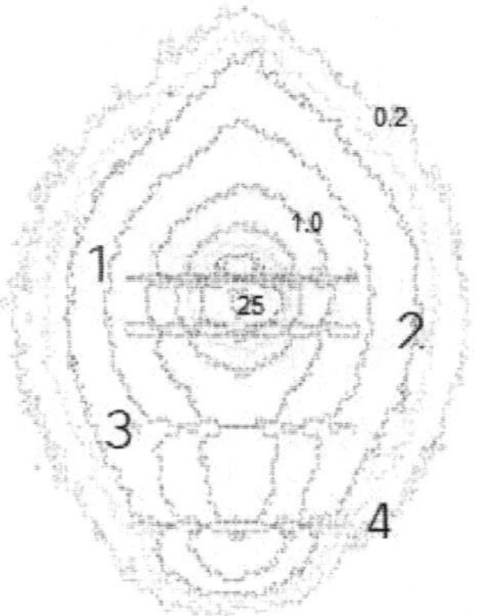

Figure 5. Processed film showing pre-programmed source.

Table1. Dwell position accuracy.

Line	Dwell Position (mm)	Relative Location (mm)	Measured Location (mm)
1	250	0	0
2	245	5	5.41
3	235	15	14.88
4	225	25	25.44

4. Discussion and future applications

Breast intraoperative radiotherapy requires a dedicated team and training in performing this procedure. Xoft eBx system comes with all necessary accessories for commisioning and implementing the system to clinical use with minimal effort. Apart from breast IORT, the system can also be used to treat skin cancer, endometrium, and cervix. Electronic Brachytherapy has rapidly grown over the last couple of years and is foreseen as a futuristic technique replacing the conventional radionuclide-based brachytherapy. Unlike the traditional radionuclide-based brachytherapy systems, the Xoft eBx is portable and can be operated in remote centres with minimal shielding. The procedure could also be performed in a mobile theatre for places where radiotherapy facilities are inaccessible. It finds wide applications in the treatment of skin, breast, spinal metastasis, rectum, endometrium and cervix. Recent clinical results for breast and endometrial cancers have confirmed excellent toxicity results due to reduced dose to organs at risk [7-9]. One of the current limitations with electronic brachytherapy systems is the source size which is an impediment to using the systems for interstitial brachytherapy. The size of a miniaturised Xoft X-ray tube is 2.25 mm in diameter, but the assembly diameter that includes the integrated water cooling sheath is 5.4 mm. The water cooling sheath enables the extraction of heat from the x-ray tube. Extraction of heat by alternate methods would reduce the size of the x-ray tube and may widen the applications of electronic brachytherapy system in radiotherapy. The diameter of a typical implant tube used in interstitial brachytherapy varies between 1.8 and 2 mm and in order to use the electronic brachytherapy, the size of the x-ray tube needs to be further reduced. Further miniaturisation of the current x-ray tube would widen the applications of electronic brachytherapy systems and also pave the way to replace radionuclide-based brachytherapy machines. Much like the lifetime of a radioactive source, the lifetime of the Xoft X-ray tube is finite. The lifetime of the tube is hardcoded to 700-900 minutes to ensure the wear and tear it is suffering is not egregious. Theoretically, an x-ray source could be stored for years between uses whereas radionuclide-based brachytherapy sources such as Ir-192 are typically exchanged every three months.

5. Conclusion

The Xoft Axxent electronic brachytherapy system has been successfully commissioned and implemented for clinical use. We find the system more user-friendly and easy to operate. It has an inbuilt dosimetry system which ensures the right dose delivered to patients undergoing brachytherapy procedures. Recent studies have confirmed the efficacy of using eBX for multiple treatment sites, and it is highly expected that eBx would replace radionuclide-based brachytherapy machines.

6. References

[1] Hurley C *et al* 2006 *Nucl. Instrum. Meth. A* **565** 801-11
[2] Baldock C 2009 *J. Phys.: Conf. Ser.* **164** 012002
[3] Healy B J *et al* 2003 *Med. Phys.* **30** 2282-91
[4] Valentin J *et al* 2005 ICRP Publication 97. *Ann ICRP* **35** 1-51
[5] Peter M *et al* 1992 United States patent US 5153900
[6] Ramachandran P *et al* 2017 *J. Med. Phys.* **42** 251-7
[7] Villamon A M *et al* 2018 *Proceedings of ESTRO 37* Barcelona, Spain
[8] Osorio J G *et al* 2018 *Proceedings of ESTRO 37* Barcelona, Spain
[9] Gandhia A *et al* 2018 *Proceedings of ESTRO 37* Barcelona, Spain

IOP Conf. Series: Journal of Physics: Conf. Series **1154** (2019) 012024 doi:10.1088/1742-6596/1154/1/012024

Dose build up characteristics with eXaSkin bolus during 6MV radiotherapy: MO*Skin* dosimetry results

T Al-sudani[1], P Metcalfe[1,2], D Cutjar[1], K Utitsarn[1], A Ceylan[2], J Davis[1] and A Rosenfeld[1]

[1]Centre for Medical Radiation Physics, University of Wollongong, Wollongong, NSW 2522, Australia.

[2]Illawarra Cancer Care Centre, Wollongong, Australia

E-mail: tahas274@uowmail.edu.au

Abstract: Skin dose evaluation is important to ensure adequate dose to the superficial target volume and to avoid severe skin toxicity. The purpose of this study is to investigate the build-up dose characteristics of eXaSkin bolus for 6MV photon beams and evaluate this material for use in skin tumor treatment. Using a 6MV Varian LINAC with field size of 10x10cm^2, the surface and build up dose characteristics of eXaSkin (thickness=3, 5, 8, 10, 15 and 26 mm) was investigated in comparison with solid water (thickness=2, 5, 15, 20 and 26 mm). Dose measurements were performed using a MO*Skin* dosimeter and advanced Markus chamber. Additionally, the percentage surface dose measurements for various oblique incident beams with 3mm thick of both bolus was investigated. Measurements were performed using MO*Skin* dosimeter and compared with advanced Markus chamber. D_{max} of eXaSkin for 6MV was approximately 0.9 cm that is consistent with the higher density of this material (ρSW=1.04 g/cm^3 ρeXa=1.7±0.03 g/cm^3). This bolus also exhibits higher dose in the build-up region between 2-10 mm depths and higher dose at oblique incident beams than solid water. Preliminary results suggest that eXaSkin is a promising candidate for use in the treatment of superficial tumors without producing excessive skin reaction.

1. Introduction

Medical linear accelerators are currently used to treat cancer with high energy X-rays (Megavoltage) [1-4]. Primary tumours are located close to many critical structures and delivering sufficient radiation dose to the primary and lymph nodes requires special attention to protect these structures [5]. As skin is critical organ, many research were investigated the immobilization devices and bolus material on skin dose [6]. Lee [7] found that the skin dose was increased about 18% caused by the bolus effect of a thermoplastic mask material during IMRT treatment of the head and neck cancer patients. Accurate assessment of surface and skin doses with eXaSkin bolus (a new material) can provide valuable information for clinical consideration to avoid near surface recurrence and severe skin toxicity.

2. Materials and Methods

2.1 Build-up region dose measurements

The eXaSkin bolus (AnatGe, Spain) is a new material (figure 1), which is composed of two components. A combined components of eXaSkin produced a new material, this material is hardening in a two minutes. For build-up measurements, the eXaSkin cuts off into 10x8 cm^2 dimensions with different thicknesses. The density of eXaSkin was calculated. The build-up region of eXaSkin at depths (0, 3, 5, 8, 10, 15 and 26 mm) was compared with build-up of solid water (Gammex, model 457, USA) (figure 2) at depths (0, 2, 5, 15, 20 and 26 mm) by using advanced Markus chamber and MO*Skin* dosimeters. PDDs were performed with 6MV Varian Clinac iX (Varian Medical Systems, Palo Alto, CA) at Illawarra Cancer Care Centre for irradiated field size of 5x5 cm^2 and 10x10 cm^2 at 100 cm SSD with 10 cm solid water as backscatter. Rawlinson [9] over-response correction formula for advanced Markus ionization chamber as follows:

$$P(\mathrm{d, E}) = P'(\mathrm{d, E, G}) - \xi(\mathrm{d, E, G}) \quad (1)$$

$$\xi(\mathrm{d, E, G}) = \xi(0, \mathrm{E, G}) \times e^{-4.0\mathrm{d}/\mathrm{dmax}} \quad (2)$$

$$\xi(0, \mathrm{E, G}) = \mathrm{c}(\mathrm{E}) \times (\mathrm{s/w}) \times \rho^{0.8} \quad (3)$$

Where P is the true PDD, P' is the measured PDD, ξ is an over-response correction factor, E is the beam energy, d is depth in the phantom, ρ is mass density of chamber wall, s/w is the ratio of electrode separation (s) to the diameter of the sidewall (w), For E=6MV photon beam energy, d_{max}=15 mm and c (E) = 0.27.

Figure 1. eXaSkin material

Figure 2. Solid water slabs phantom

2.2 oblique incident beam

The percentage surface doses (PSD) for different incident angles were measured using an advanced Markus chamber and MOSkin. The oblique incidence beam was investigated at 0°, 15°, 30°, 45°, 60° and 75° beam angles for 6 MV photon beams, 100 SSD and 10x10 cm^2 field size. Furthermore, the effect of different incident beam angles on the 3mm solid water and 3mm eXaSkin bolus was compared.

3. Results and Discussion

3.1 build-up region dose measurements

The density of eXaSkin material was calculated by measure its slabs dimensions and their weights. The density of eXaSkin is 1.7± 0.03 g/cm^3 compared with the solid water, which it is, equals to 1.04 g/cm^3. (Figure 3 and figure 4) show the PDD measurement by the advanced Markus chamber and MOSkin dosimeter with the eXaSkin and the solid water phantom as a bolus material for field size of 5x5 cm^2 and 10x10 cm^2. The PDD of eXaSkin for both field sizes are higher than PDD of solid water at (2-10 mm).

The d_{max} of solid water was at 15mm depth (density 1.04 g/cm^3) and when compared it with the eXaSkin density (1.7 g/cm^3), the d_{max} of eXaSkin was 9 mm.

The comparison of the PDD of eXaSkin between advanced Markus measurements and MO*Skin* measurements for both field sizes 5x5 and 10x10 cm^2 (Figure 5) shows that a large difference is at 0mm for both field sizes (7%) and that attribute to the water equivalent depth (WED of MO*Skin* is 0.07 mm while WED of advanced Markus 0 mm after applying over response correction factor).

IOP Conf. Series: Journal of Physics: Conf. Series **1154** (2019) 012024 doi:10.1088/1742-6596/1154/1/012024

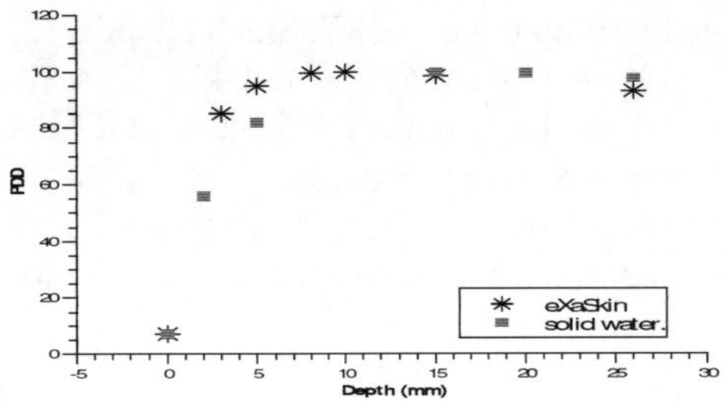

(a)

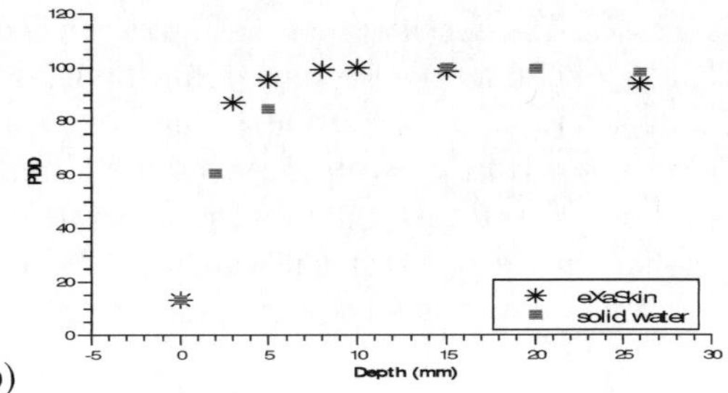

(b)

Figure 3. Build up region for solid water vs eXaSkin by using advanced Markus chamber for a) 5x5 cm^2 and b) 10x10 cm^2.

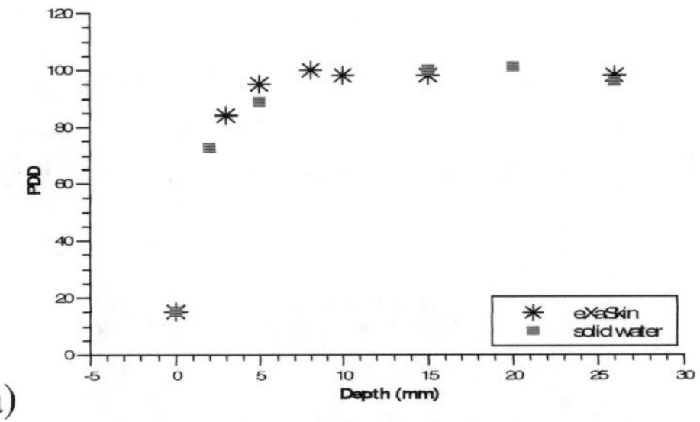

(a)

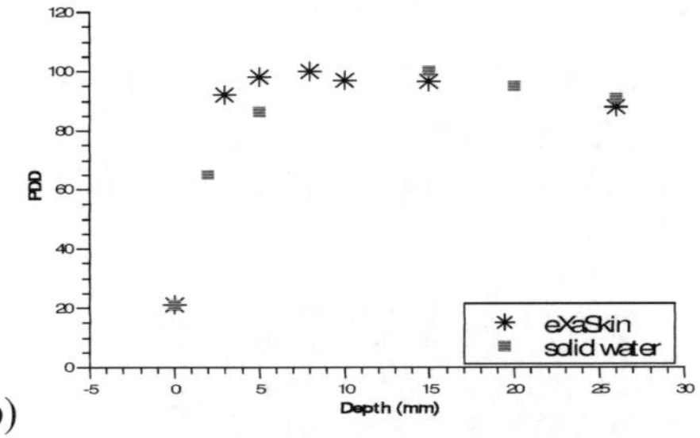

(b)

Figure 4. Build up region for solid water vs eXaSkin by using MO*Skin* dosimeter for a) 5x5 cm^2 and b) 10x10 cm^2.

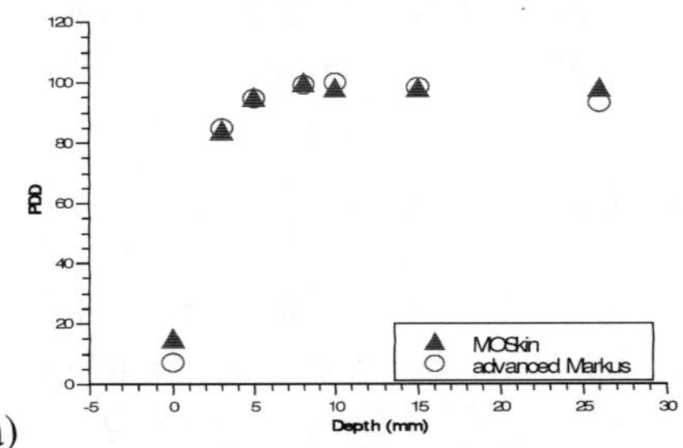

(a)

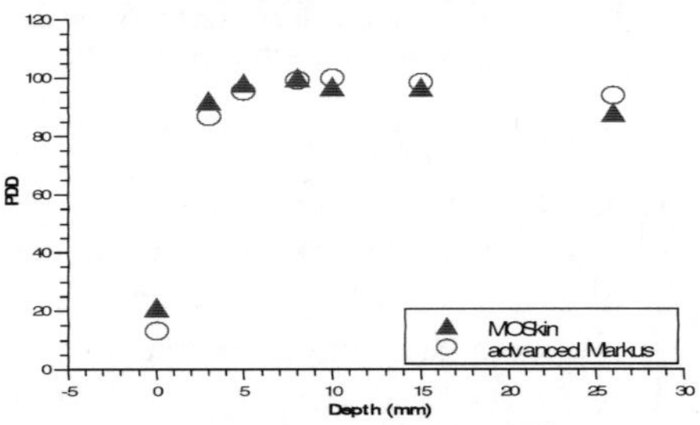

(b)

Figure 5. Comparing the PDD of eXaSkin in the build-up region between advanced Markus measurements and MOSkin dosimeter for a) 5x5 cm^2 and b) 10x10 cm^2.

3.2 oblique incident beam

Figure 6a demonstrates the alteration of the percentage surface dose with oblique incident beam effect for 6MV, 100SSD and 10x10 cm^2 field size. Advanced Markus uncorrected measurements and MO*Skin* are used. Percentage dose increases when incident beam angle increases because the charge particle equilibrium CPE shifts near the surface and that agree with Jong [8] measurements. The measurements of percentage dose with 3mm of eXaSkin are the highest as in (figure 6c) when compared it with the percentage dose at 3 mm solid water (figure 6b) and at surface (figure 6a). The large difference in measurements between advanced Markus chamber and MO*Skin* with 3mm eXaSkin is -11% at 75° that attribute the eXaSkin bolus not covered the entire irradiated surface.

IOP Conf. Series: Journal of Physics: Conf. Series **1154** (2019) 012024 doi:10.1088/1742-6596/1154/1/012024

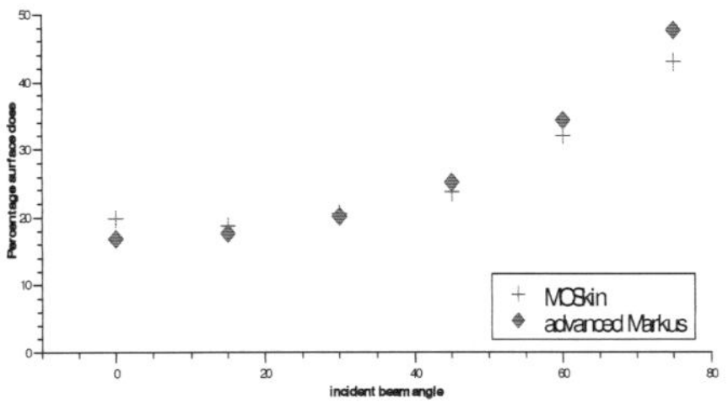

(a)

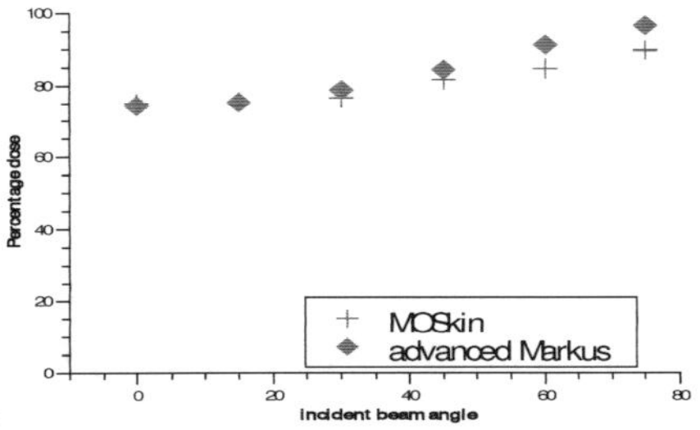

(b)

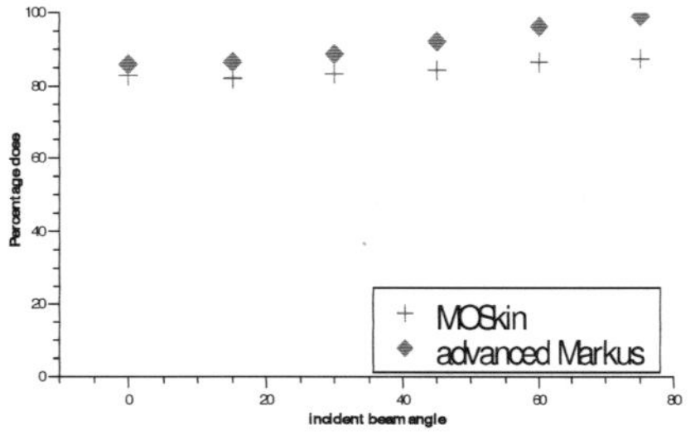

(c)

Figure 6. Percentage dose for the oblique incident beam by using advanced Markus chamber and MO*Skin* dosimeter a) at surface b) at 3mm solid water and c) at 3mm eXaSkin.

4. Conclusion

Skin dose assessment with eXaSkin bolus is essential to ensure deliver dose to the tumors near surface without producing excessive skin reaction. The new bolus exhibits high dose at depths 2-10 mm compared with solid water. It also exhibits high dose with different angle incident beams. This bolus is a viable option to treat the tumours near the surface.

5. Acknowledgments

We would like to thank Sarah Neylon (NL-Tec Pty Ltd (Sydney)) for providing us eXaSkin bolus.

6. References

[1] Vial P *et al* 2006 *Phys. Med. Biol.* **51** 5517-38
[2] Vial P *et al* 2008 *Med. Phys.* **35** 4362-74
[3] Gustafsson H *et al* 2009 *Med. Phys.* **36** 5665-74
[4] Venning A *et al* 2005 *Nucl. Instrum. Meth. A* **555** 396-402
[5] Shang Q *et al* 2015 *Appl. Radiat. Oncol.* **9**
[6] Hsu S H *et al* 2008 *Phys. Med. Biol.* **53** 2593-606
[7] Lee N *et al* 2002 *Int. J. Radiat. Oncol. Biol. Phys.* **53** 630-7
[8] Jong W L *et al* 2014 *J. Appl. Clin. Med. Phys.* **15** 120-32
[9] Rawlinson J A *et al* 1992 *Med. Phys.* **19** 641-648

IOP Conf. Series: Journal of Physics: Conf. Series **1154** (2019) 012025 doi:10.1088/1742-6596/1154/1/012025

Modelling the x-ray source for the Australian MRI-Linac

N Roberts[1,2], B Oborn[1,3], U Jelen[2], B Dong[2], J Begg[2,4,5], A George[4], S J Alnaghy[1,2], T Causer[1,2,3], T Alharthi[2,6], L Holloway[1,2,3,4,5] and P Metcalfe[1,2]
[1]Centre of Medical and Radiation Physics, University of Wollongong, NSW, Australia
[2]Ingham Institute for Applied Medical Research, Sydney, NSW, Australia
[3]Illawarra Cancer Care Centre, Wollongong Hospital, Wollongong, NSW, Australia
[4]Department of Medical Physics, Liverpool and Macarther Cancer Therapy Centre, Sydney, NSW, Australia
[5]South Western Sydney Clinical School, University of New South Wales, Liverpool, NSW, Australia
[6]Institute of Medical Physics, School of Physics, University of Sydney, Sydney, NSW, Australia

E-mail: nr593@uowmail.edu.au

Abstract. MRI-guided radiotherapy allows real-time imaging during treatment however the magnetic field influences the dose distribution in the patient. An accurate model of the radiation beam and the encompassing magnetic field is important to predict dosimetry changes. The purpose of this work is to develop a Monte Carlo model of the Australian MRI-Linac to be used as input into a dose calculation tool for treatment planning. The Australian MRI-Linac is a 1 T inline system with a 6MV flattening filter free photon beam. Commissioning measurements were undertaken both with and without the magnetic field present, PDDs and profiles were used to develop a model with the Geant4 toolkit. To date the model at 0 T matches within ±2% of measured data. Ongoing work involves measurements at 1 T at various linac to MRI isocentre distances, the magnetic field model at each configuration is also under development.

1. Introduction

MR-guided radiotherapy offers superior soft tissue contrast and the potential to reduce current treatment margins while escalating dose to the tumour [1]. The radiotherapy beam can be adapted to deliver a more precise treatment by utilising knowledge of tumour position and shape obtained from the MRI in real time. While there are many potential advantages to this treatment technique, there are still many engineering and physical challenges which must be resolved. Understanding how the magnetic field produced by the MRI impacts dose distributions inside the patient is something which can be modelled using Monte Carlo methods. For the Australian MRI-linac [2] it is important to include the fringe field of the MRI in calculations due to the increased electron hot spot at the surface caused by this particular field [3,4]. Since current commercial treatment planning systems are not capable of modelling the complete magnetic field, including the fringe field, we have proposed a Monte Carlo dose verification system to calculate dose with and without the magnetic field. The focus of this work is to develop a model of the x-ray beam for such a system for the Australian MRI-linac.

2. Materials and Method

The current Australian MRI-linac prototype is a 1 T split-bore MRI (Agilent Technologies, Oxford UK) with a 6MV FFF horizontal photon beam (Varian Linatron-MP), and MLC system (Varian Millennium 120-leaf). The linac and MLC are mounted on a rail system so that source to imaging isocentre distance can be varied between 1.8-3.3 m.

Initial commissioning of the linear accelerator was performed prior to installation of the MRI. These measurements were acquired under 0 T conditions, and used to assist in developing an accurate model of electron beam parameters for the linear accelerator. The Monte Carlo toolkit, Geant4 [5] version 10.2.p01 was used to model the beam to obtain a close match to experiments. PDDs and profiles for the 6MV beam were taken in a water tank with a CC13 ion chamber and in solid water with Gafchromic EBT3 film. The model includes the linear accelerator treatment head, MLC and phantoms used during measurements as shown in figure 1a. Mean energy, FWHM of the energy distribution and focal spot size of the incident electron beam were optimized in the simulation to match experimental results. Measurements were taken at extended source to isocentre distances (SIDs), corresponding to the adjustable SIDs following installation of the MRI. This variation in source position means the field sizes are also variable, to keep consistent all field sizes are quoted as input into the MLC controller defined as the field size at 1 m from the source.

Commissioning measurements after installation of the MRI were performed with EBT3 film in solid water, profile shapes were cross checked with the STARCHECK$^{maxi\ MR}$ (PTW, Freiburg, Germany), all measurements were taken at an SID of 2.869 m. The Agilent MRI magnet was modelled in COMSOL Multiphysics (Stockholm, Sweden) to produce a 3D magnetic field map. Figure 1b shows simulation geometry with the MRI model displaying its cross section to visualise the coils.

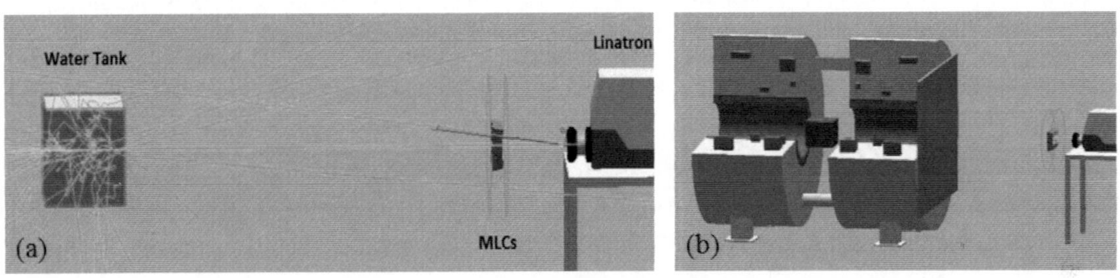

Figure 1. Geant4 model of Australian MRI-linac, 0 T setup which includes the linac, MLC and water tank (a) 1 T setup with MRI, RF panel, linac, MLC and solid water phantom at isocentre (b)

IOP Conf. Series: Journal of Physics: Conf. Series **1154** (2019) 012025 doi:10.1088/1742-6596/1154/1/012025

3. Results

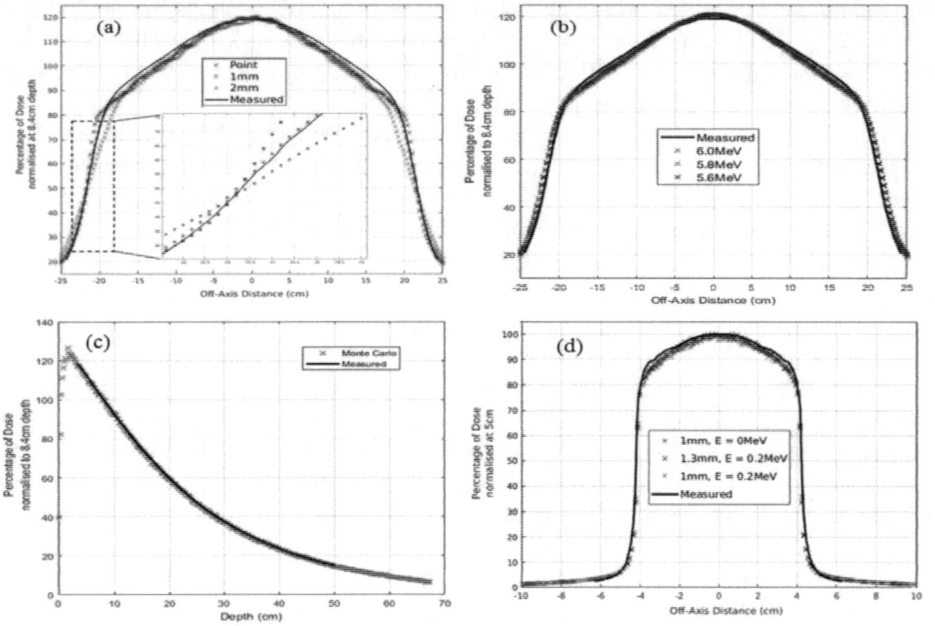

Figure 2. Comparison of measured (black) and simulated data (crosses). Open field at SID 1.9 m (a)-(c) variations in spot size of the incident electron beam (a) variations in energy of the incident electron beam (b) PDD for energy of 6 MeV and spot size of 1 mm (c). 10x10 cm^2 field at SSD 0.743 m, variation in energy spread and spot size (d)

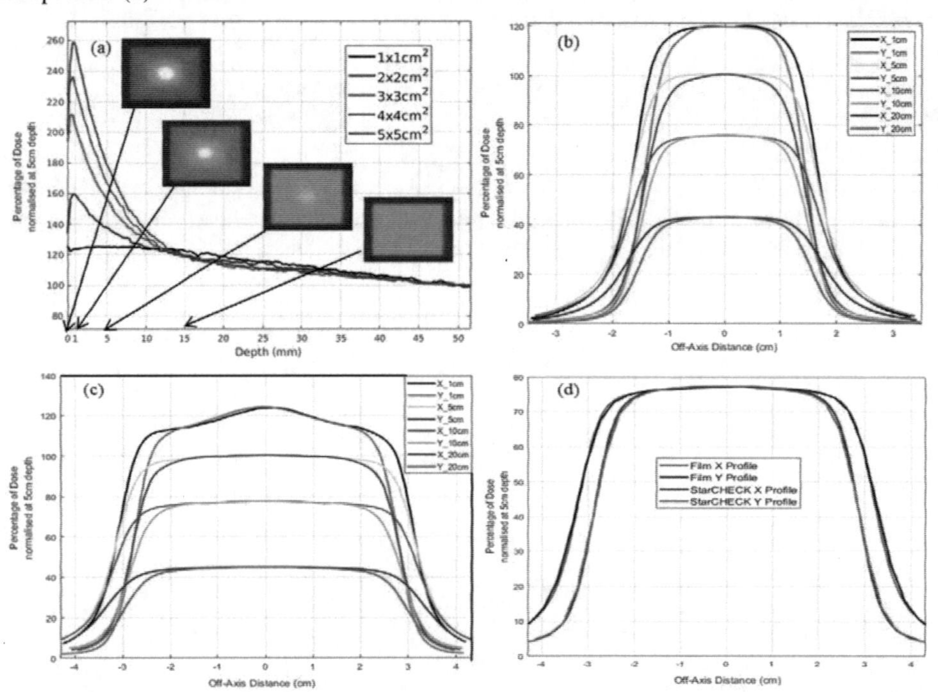

Figure 3. Measured data post MRI installation. PDDs for different field sizes, 2D dose distributions from 3x3 cm^2 field simulations overlaid (a), Profiles at different depths for 1x1cm^2 (b) and 2x2 cm^2 field (c), Comparison of film and STARCHECK$^{maxi\ MR}$ measurements at depth 10 cm for 2x2 cm^2 field (d)

4. Discussion

Open field data was collected for the initial 0 T commissioning measurements, comparing with simulations, with an incident electron beam of spot size 1mm the match of the beam penumbra is closest (figure 2a).

Energies from 5.6-6 MeV yielded similar PDDs and profiles for this SID (figure 2b, 2c). To investigate the spot size further, we took measurements at 0.743 m SSD. Varying the simulation spot size and energy spread at this SSD had minimal effects on the profiles (figure 2d). Since the SID will not be less than 1.8 m for the current system, the field sizes of interest would be 10x10 cm^2 (at 1m) or smaller, therefore the current match for 0 T data 10 cm off-axis at isocentre is sufficient.

The effects of the magnetic field on the dose distribution are dependent upon the orientation of B_0 field with respect to beam direction. In the current system the beam is parallel to the magnetic field; an advantage of this is reduced effects of the Lorentz force, however due to the magnet's fringe field we see an increased dose to the surface due to contaminant electrons being focused along the central axis [3,4]. This effect can be seen in measured data shown in figure 3a, the larger the field size the higher the percentage dose at the surface. Overlaid, on figure 3a, are 2D dose distributions from simulations where the electron hot spot can be seen through the first few millimetres of the phantom, these electrons have all been absorbed by 1.5 cm depth. Profiles taken with film for field sizes 1x1 cm^2 and 2x2 cm^2 are also shown (figure 3b, 3c). A comparison of the 2x2 cm^2 field profiles taken with film and the STARCHECK$^{maxi\ MR}$ (figure 3d) are in agreement, validating our film dosimetry methods.

5. Conclusion

A Monte Carlo model of the x-ray beam for the Australian MRI-linac system has been developed. Agreement within ±2% between modelled and measured PDDs and beam profiles was observed at 0 T. Commissioning measurements in the presence of the 1 T field have been performed and cross checked for a specific SID. Future work will involve similar measurements with the linac at shorter SIDs and further development on the magnetic field model at each SID.

7. Acknowledgements

The authors acknowledge Cancer Council NSW Project Grant (APP1128336) and NHMRC Program Grant (APP1036078). The author N. R. receives scholarship support from the Australian Government Research Training Program Scholarship and the Centre for Oncology Education and Research Translation.

8. Reference

[1] Menten M *et al* 2017 *Physica Medica.* **44** 139-149
[2] Keall P J *et al* 2014 *Semin. Radiat. Oncol.* **24** 203-6
[3] Oborn B M *et al* 201, *Med. Phys.* **39** 879-890
[4] Oborn B M *et al* 2014 *Med. Phys.* **41** 051708
[5] Agostinelli S *et al* 2003 *Nucl. Instumen. Meth. A* **506** 250-303

Preliminary epi-diode characterization for HDR brachytherapy quality assurance

A A Romanyukha, M Carrara[2], D Cutajar[1], M Petasecca[1], I Fuduli[1], T Al-Salmani[1], M.Lerch[1], V.Perevertaylo[3], D Mazzeo[2], A Cerrotta[2], C Fallai[2], E Pignoli[2] and A B Rosenfeld[1]

[1]Centre for Medical Radiation Physics, University of Wollongong, Wollongong, Australia

[2]Medical Physics Unit, Department of Diagnostic Imaging and Radiotherapy, Istituto Nazionale dei Tumori, Milan, Italy

[3]SPA BIT, Kiev, Ukraine

E-mail: anatoly@uow.edu.au

Abstract. High Dose Rate vaginal brachytherapy for endometrial cancer has evolved from simple single-channel (i.e. cylindrical applicator) deliveries to treatments involving several channels (i.e. multichannel applicator) for the radiotherapy source to dwell, increasing the complexity of the dose distribution, and allowing more space for potential errors. For this reason real-time treatment verification has gained a greater importance than ever before, and more methods need to be developed in order to provide assurance that the dose delivery has been carried out as intended by the hospital staff. P-type silicon epi diodes have been designed at the Centre for Medical Radiation Physics (CMRP) in Australia to suit the specific needs of HDR BT, and characterized in the clinical BT facility of the Fondazione IRCCS (INT) in Italy. They have shown great potential for BT treatment verification in real time due to their radiation hardness, dose rate independence, flexibility in physical design, and ability to monitor the treatment at a 1-kHz readout frequency. Their dynamic range has been determined as ±17 to ±20 mm and dwell time calculation accuracy of > 0.1 s has been shown. If placed on the same longitudinal plane of a treatment accessory, these detectors would enable coverage of about 40 mm for source position and dwell time tracking. Respective detector positioning at (0, +3, -3 mm) would extend this range to 45-50 mm, depending on the catheter location, proving to be sufficient for the majority of treatment cases. Further studies are encouraged to develop diodes with a wider dynamic range.

1. Introduction

High dose rate (HDR) brachytherapy (BT) is a radiotherapy (RT) technique that is characterized by its ability to deliver a highly conformal dose to the target volume, and often delivers high doses per fraction that have a steep dose gradient [1]. Disease sites treated with HDR BT include, but are not limited to breast, prostate, and the gynaecological cavity. In the simplest gynaecological treatments the HDR source is administered using a cylindrical applicator with a single central channel for the source to dwell in a range of positions and dwell times, and can be customized based on the patient's needs.

The single-channel applicator has evolved into a multi-channel applicator, allowing increased flexibility in the dose distribution [2]. The multi-channel applicator contains one central channel, often with the ability of inserting an intra-uterine tube, and several equally spaced peripheral channels.

Flexibility in the treatment also leads to higher complexity and thus presents more room for error, for example transfer tube connections may be accidentally switched during treatment setup in the case of multiple applicator channels. For this reason treatment verification is desirable, especially one that is available in real-time and allows the treatment to be interrupted in case an error is detected.

A number of radiation detectors have been investigated for treatment verification [3, 4], both for *in vivo* dosimetry [5-8] and HDR source tracking purposes [9-11]. Source tracking in particular has been previously achieved using a number of imaging techniques, such as pinhole cameras [12, 13] and EPID imaging [14-17]. Our aim is to develop a system that is able to verify both positions and dwell times in real-time. A comprehensive treatment verification system that employs the tracking of the HDR source is one that is able to determine both the positions and dwell times of the source, is capable of real-time readout, and allows a high spatial and temporal resolution comparable to that of HDR afterloaders used to administer the source. Radiation detectors used in such a system must therefore comply with the aforementioned description.

The objective of this study is to evaluate the suitability of epi diodes that have been developed at the Centre for Medical Radiation Physics (CMRP), and evaluate their sensitivity with respect to measurement settings, source air-kerma strength (S_K), and source-to-detector distance, as well as to establish a method for source dwell time verification.

2. Materials & Methods

The most commonly used source for HDR BT treatments is currently Ir-192, which has a half-life of ~74 days and is used for treatment in the clinic for a period of three months. During this timeframe the source strength (S_K) normally ranges between approximately 10 and 45 $mGy.m^2.h^{-1}$. At the Fondazione IRCCS Istituto Nazionale dei Tumori (INT) in Milan, Italy, brachytherapy is administered using the microSelectron afterloader unit, while the patient-specific treatments are planned using the Oncentra Treatment Planning System (TPS).

P-type epi diodes have been developed at the CMRP in the University of Wollongong to suit the specific needs of HDR BT quality assurance. They measure 325 mm in length and contain a sensitive silicon chip of 1.5 x 1.5 x 0.038 mm^3 (fig. 1a,b). The detectors are tailored to treatment verification in the form of real-time source tracking and possess features such as radiation hardness, dose rate independence, high readout frequency, and an un-biased readout guaranteeing suitability for in-patient use. The readout system consists of three boards: the AFE (analog front end), the FPGA (field programmable array), and the power supply board. The reader has four channels, and was covered in aluminium to minimize background noise (fig. 1c). The FPGA is connected via a USB connection to a laptop with CMRP-developed readout software AFE-Histogram that displays the instantaneous and integral responses of the four available channels. Readout settings can be adjusted to suit the particular type of radiation source and measurement setup, and include the range, integration time, and frequency of readout. The range indicates the number of times the capacitor has discharged, and can be set between 0 and 7 to reflect the AFE sensitivity. Integration time is the window of the AFE signal integration, and can range between 50 and 400 μs, whereas the frequency of readout can range between 1 and 1000 Hz.

Current-voltage (I-V) characteristics of the diodes were verified at the start of the study to ensure a stable relationship between applied voltage and current within the diodes by testing them in reverse biased mode prior to connector and shielding assembly.

The detectors were then positioned inside a water equivalent phantom at a distance of approximately 3 mm from the source on the longitudinal axis, approximately in the middle of the 48 available dwell positions. The HDR source was sent to all 48 available positions with the default step size of 2.5 mm and dwell times of 2 seconds. Diode response was acquired as the charge collected in the diode sensitive volume (pC). The range, acquisition time, and frequency of readout were modified between experiments to determine the optimal setup for source tracking in real-time.

To ensure diode dose rate independence and evaluate the variability of its response with respect to source activity throughout the lifetime of the source, measurements were acquired at different Ir-

IOP Conf. Series: Journal of Physics: Conf. Series **1154** (2019) 012026 doi:10.1088/1742-6596/1154/1/012026

192 source strengths. Diode dynamic range was investigated in terms of dwell position distance from the centre of the sensitive volume and assumed to extend until 95% of the readout points no longer conformed to their corresponding cluster group i.e. dwell position response range (pC), within two significant figures. A method of dwell time verification was also determined.

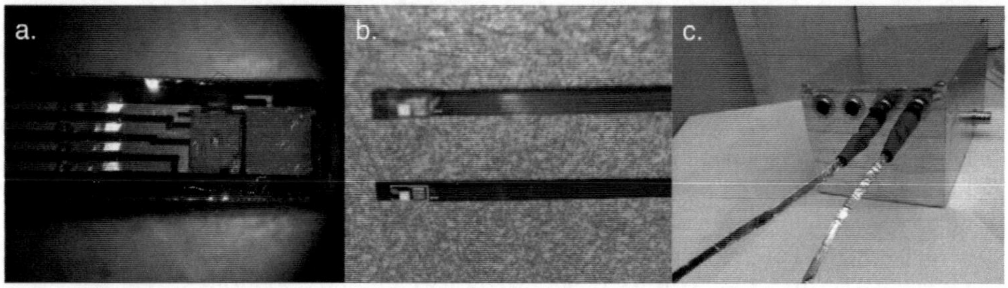

Figure 1. (1a) Diode bonds and silicon sensitive chip (1.5 x 1.5 x 0.038 mm^3 as viewed under a microscope). The bonds were covered for protection (b), and the complete diode assembly is used in conjunction with the in-house produced reader (c).

3. Results and discussion

I-V tests have indicated good P-N junction operation of all three selected dosimeters i.e. D1, D2, D3 (fig. 2). Three-pin female connectors were then attached to the detector pigtails and aluminium thin foil shielding added above the Kapton tail to minimise noise arising from the electronics and light exposure.

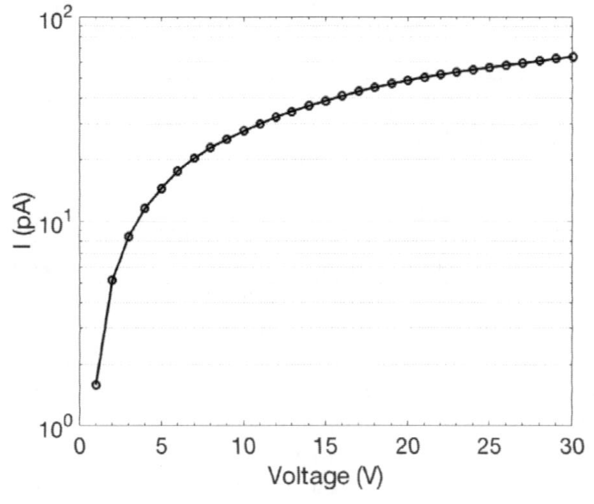

Figure 2. Typical current-voltage characteristics of the diode set.

Integration time and readout frequency were established as 400 μs and 1 kHz, respectively, to allow the diode to be read out at its maximum sensitivity. Fig. 3 shows S_K-normalised mean diode response as a function of treatment time for range (sensitivity) settings of 4, 5, and 6 and S_K of 25.4 $mGy.m^2.h^{-1}$. The response peaks while the source is closest to the diode's sensitive volume, and gradually decreases as the source moves away in either direction. At ranges 4 and 5 oversaturation occurs before the source is able to reach the closest dwell position, while a range of 6 allows the response to reach its maximum. A range setting of 6 or higher was therefore determined as optimal for measurement with the Ir-192 source to accommodate the range of dose rates throughout the lifetime of the HDR source in the hospital environment.

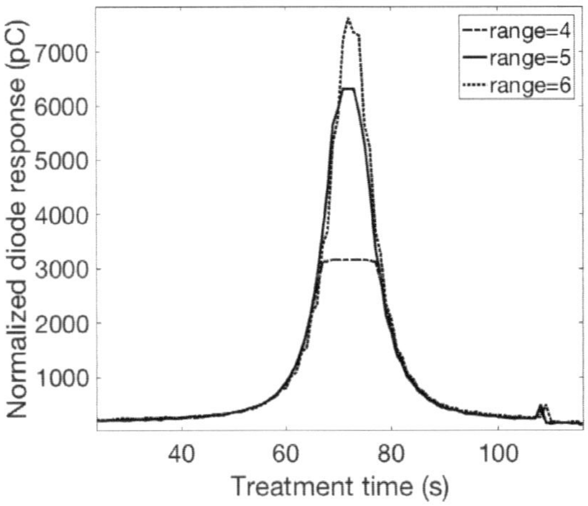

Figure 3. S_K-normalized mean diode response in each dwell position for range settings of 4, 5, and 6.

Diode response as a function of source strength, with S_K = 38.0, 34.6, 13.2 mGy.m^2.h^{-1} is shown in fig. 4a, where the highest range of measured charge plotted in black corresponds to the highest S_K value of 38.0 mGy.m^2.h^{-1}. The same three curves normalized for S_K demonstrate some dependence on the source dose rate (fig. 4b), especially when dealing with the maximum and minimum S_K values used in the clinic. Therefore, in order to increase the accuracy of source localisation, epi diodes should be calibrated at three different S_K strengths within the clinically relevant range, and expected diode response as a function of source position for individual catheters should be adjusted by means of interpolation, and embedded in a software look up table.

Moreover, due to a decrease of the response at lower dose rates diode dynamic range slightly decreases at lower S_K (diode response for S_K=13.23 is plotted in blue in fig. 4b), as shown by the rise in tail noise i.e. dwell positions located further away. It appears that normalised diode response is of similar magnitude for source positions of ±12.5 mm, while the signal-to-noise ratio is reduced at lower source activities and the response changes for positions located further away. The response peak at a treatment time of ~115 seconds corresponds to the source retraction into the afterloader, which is important for verification of its performance.

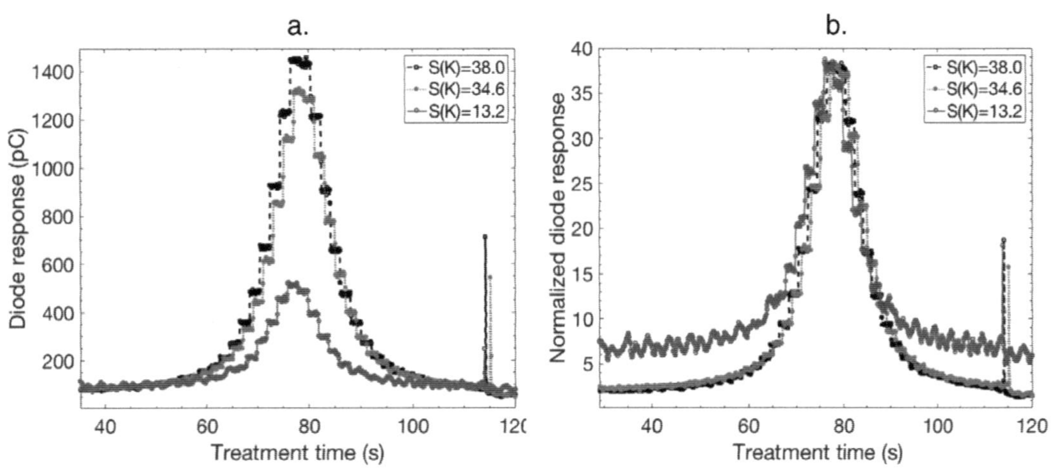

Figure 4. Diode response for various source strengths: total diode response (4a), and S_K-normalised response (4b).

Normalised mean diode response i.e. R_{DIODE} at individual dwell positions is plotted as a function of source distance from the centre of the diode's sensitive volume (mm) in fig. 5. Diode dynamic range was determined as ±20 mm, ±18 mm, and ±17 mm for diodes 1, 2, and 3, respectively. The fluctuation in diode response in a single dwell position was 0.5 - 2.6% for diode 1, 0.3 - 1.3% for diode 2, and 0.9 - 3.0% for diode 3, with higher fluctuation at increased source-to-detector distances. The diode range established in this study indicates that the combination of the three diodes would enable the coverage of approximately 40 mm of a gynaecological applicator, which may in some cases be shorter than the region of interest in HDR vaginal treatments for endometrial cancer. However if diodes are placed at (0, +3, -3 mm, where 0 is the middle position of the expected clinical range of source dwelling) of one another, then a coverage of the top 45-50 mm of the applicator would be possible, and would prove sufficient for the majority of treatment cases.

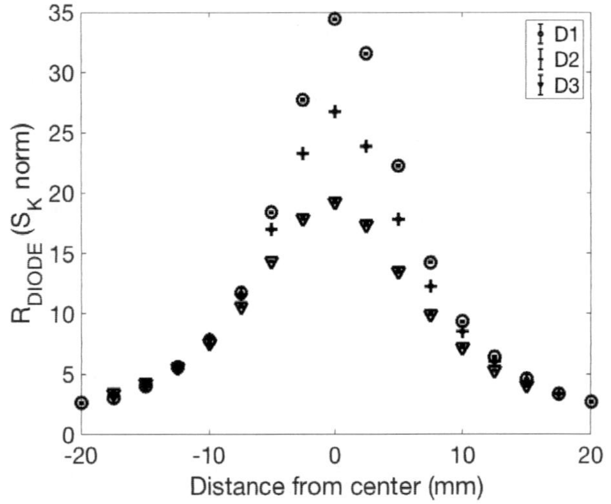

Figure 5. Normalised mean diode response in each dwell position as a function of the dynamic range of each diode.

Dwell time was determined from the time of constancy of diode response, and is represented by the signal plateaus in fig. 6. Dwell time is calculated as the product of the number of measurements in each plateau and the readout time interval, while readout signal values are constant. Measured dwell time was verified against the nominal dwell time with an accuracy of ± 0.3 s averaged over all dwelling positions. As the readout frequency was set below 10 ms, source transit time can in some cases be observed in fig. 6 as the single points between the plateaus.

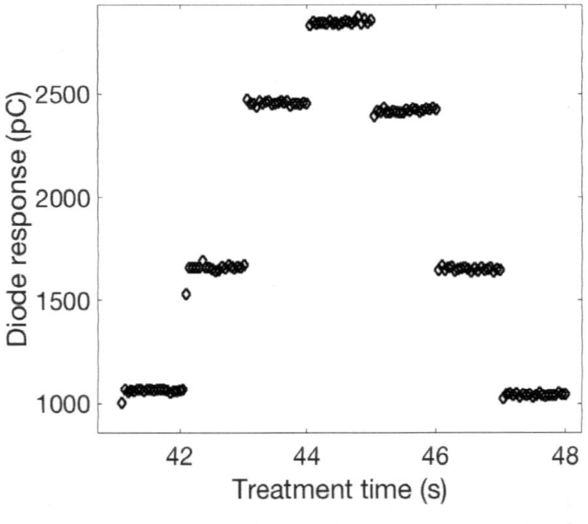

Figure 6. Diode response as a function of treatment time, where each response plateaus corresponds to individual dwell positions of 2 seconds in duration.

Source transit is visualised in fig. 7 before and after the source reaches a stable dwell position, traveling past the diode in both directions. The source transit time was calculated as 0.3 s both as it travelled from the afterloader to the planned dwell position and as it retracted back (shown in red), for a total of 0.6 s.

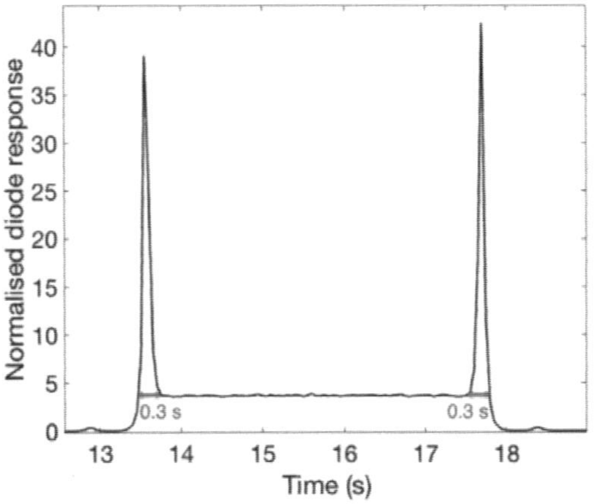

Figure 7. Diode response peaks corresponding to the source transit.

At the 0.001 s readout frequency the diodes have thus shown the ability to verify source dwell time with a better than 0.1 s accuracy in each position. The higher discrepancies in measured dwelling time determined in comparison to the TPS are attributed to source transit, equating to ~0.6 seconds per dwell position. This is taken into account in the TPS dose calculation and leads to lower physical dwell times than indicated by the TPS or measured experimentally, as above, which explains the ± 0.3 s accuracy.

4. Conclusions

The characteristics and suitability of epi diodes developed for HDR source tracking has been studied. Diode response to the HDR BT Ir-192 source has been evaluated with respect to measurement settings, source-to-detector distance, and source air kerma strength. Optimal readout settings allow a balance between minimal noise and maximum diode sensitivity to Ir-192, and are especially important in the clinical environment due to the wide range of S_K of an individual source; the dynamic range of diodes has been defined and determined adequate for use in HDR BT full treatment verification, if three detectors are placed approximately 3 mm apart. A method for dwell time calculation has been developed and tested to provide a better than 0.1 s accuracy, and a ±0.3 s mean accuracy as compared to the TPS. Although time resolution is appropriate for clinical needs, the dynamic range may in some cases be shorter than the relevant region in HDR vaginal treatments for endometrial cancer. The diode system has shown great potential, and further studies are encouraged to develop more sensitive diodes capable of detecting the source position over a wider dynamic range.

5. Acknowledgments

We thank the Australian Research Council for scholarship support of A.A.R. through Laureate Fellowship grant FL130100116. This work was partially supported by Italian Ministry of Health (5 x 1000 Funds - 2014).

6. References

[1] Hurley C *et al* 2006 *Nucl. Instrum. Meth. A* **565** 801-11
[2] Carrara M *et al* 2017 *Physica Medica* **44** 58-65
[3] Carrara M *et al* 2017 *Brachytherapy* 17 133-45
[4] Kertzscher G *et al* 2014 *Br. J. Radiol.* **87** 20140206
[5] Tanderup K *et al* 2013 In vivo dosimetry in brachytherapy. *Med. phys.* **40** 070902

[6] Carrara M *et al* 2017 *Phys. Med.* **41** 5-12
[7] Carrara M *et al* 2016 *Radiother. Oncol.* **118** 148-53
[8] Lambert J *et al* 2007 *Med. Phys.* **34** 1759-65
[9] Therriault-Proulx F *et al Med. Phys.* **38** 2542-51
[10] Nakano T *et al* 2003 *Phys. Med. Biol.* **48** 2133-46
[11] Smith R *et al* 2013 *Med. Phys.* **40** 111706
[12] Safavi-Naeini M *et al* 2015 *Med. Phys.* **42** 7098-107
[13] Duan J *et al* 2001 *Med. Phys.* **28** 167-73
[14] Smith R *et al* 2013 *Med. Phys.* **40** 12 111706
[15] Fonseca G *et al* 2017 *Phys. Med. Biol.* **13** 5440-61
[16] Vial P *et al* 2008 *Med. Phys.* **35** 4362-74
[17] Gustafsson H *et al* 2009 *Med. Phys.* **36** 5665-74

IOP Conf. Series: Journal of Physics: Conf. Series **1154** (2019) 012027 doi:10.1088/1742-6596/1154/1/012027

Characterisations of a fibre optic dosimetry system for source tracking during HDR Brachytherapy

M Al Towairqi[1,2,3], D L Cutajar[1], T Braddock[1], E Li[1], S Wadi-Ramahi[3], B Moftah[3,4] and A B Rosenfeld[1]

[1]School of Physics, Centre for Medical Radiation Physics, University of Wollongong, Northfields Ave, Wollongong NSW 2522, Australia

[2]School of Physics, Faculty of Science, Taif University, Airport Road, Al Hawiyah, At Taif 26571, Saudi Arabia

[3]Biomedical Physics Department, King Faisal Specialist Hospital& Research Centre, Zahrawi St, Al Maather, Riyadh 12713, Saudi Arabia

[4]Medical Physics Unit, McGill University, 845 Sherbrooke St W, Montreal, QC H3A 0G4, Canada

Email: mhat989@uowmail.edu.au

Abstract. Brachytherapy is a complex treatment procedure where radioactive sources are inserted in or close to the tumours to destroy the cancerous cells. Due to the unique properties of scintillation plastic detectors, this study was aimed to characterise an innovative fibre optic dosimetry system as a quality assurance tool during HDR Brachytherapy. Scintillating plastic fibres with different scintillation lengths were prepared and then optically coupled to non-scintillating fibres for light transmission. A transimpedance photodiode amplifier was used to detect positional sensitivities of different fibre probes placed within a solid-water phantom at varying distances above an 192-Ir brachytherapy source located within the catheter. Monte Carlo simulation has validated the expected response of the scintillating plastic fibres for multiple dwell positions with demonstrating the variance of detector response with source location. It showed the ability of shorter scintillating fibre lengths to distinguish between varying source locations when the SNR maintained high. However, fully scintillating plastic fibre showed flat response for most dwell positions. The proposed system proved to be appropriate for further clinical investigations, such as simultaneous dose measurement, and providing 3D position reconstruction through in-vivo source tracking.

1. Introduction

Brachytherapy is a complex radiotherapy technique where radioactive sources, with a high dose rate usually 12Gy/h and more, are inserted close to or within the tumour volume to treat cancerous cells. Hence, quality assurance (QA) programs have been established worldwide by introducing sophisticated QA devices for assuring the safe delivery of radiation dose to patients, verifying and tracking HDR source during Brachytherapy treatment to prevent unexpected errors that might lead to disastrous accidents in clinical scenarios [1-4]. Due to the desired properties, such as water equivalence, high spatial resolution, and dose rate independence, that fibre optic scintillation detectors possess compared to other conventional QA tools, they have shown the capability for effectively performing *in-vivo* dosimetry [5]. However, the only disadvantage of such technology is being nearly temperature dependant [6], and also the production of Cherenkov light when charged particles travel at

IOP Conf. Series: Journal of Physics: Conf. Series **1154** (2019) 012027 doi:10.1088/1742-6596/1154/1/012027

a speed higher than the speed of light in that medium [7], which is considered to be a major source of unwanted signals that need to be corrected for when measuring the dose.

There has been a considerable amount of research on fibre optic dosimetry (FOD) as real-time quality assurance tools in radiotherapy. The continuing growth and the unique properties have undoubtedly attracted the attention of researchers which resulted in a significant number of papers published. At the end of the last century, in 1992, Beddar *et al* introduced the first scintillation fibre optic dosimetry system that resulted in the creation of a scientific revolution in the path of radiation detection and measurements which opened up wide horizons for those who are interested in FOD [8, 9]. However, several research groups around the world have applied this fabulous technology for *in-vivo* dosimetry with so many different radiation modalities. It is worth mentioning the successful implementation of FOD with photon [9-12], electron [13, 14], proton beams [15, 16], radiology [17, 18], and Brachytherapy [19-21]. This study aims to introduce an innovative, portable, cost effective fibre optic system as a QA tool, developed to measure the dose and localise the HDR Brachytherapy source based on the response of multiple fibres in catheters within the treatment volume. In addition, it also aimed to characterise the system in terms of the optical properties and selecting optimal fibre lengths and distance to the Ir-192 source.

2. Materials and Methods

2.1 Preparation of fibre optic probes

Scintillating plastic fibres (BCF-60, Saint-Gobain Crystals USA) with different scintillation lengths (from 10cm to 0.5cm) were in house-prepared and tested before being used. All fibres were finely cut using a sharp cutter, polished using polishing films of different thickness, paint coated, and then optically coupled to a 1m long clear plastic fibres (BCF-98, Saint-Gobain Crystals USA).

2.2 Experimental work

A Monte Carlo simulation was performed using the Geant4 toolkit (4.9.2) to validate the methodology behind converting signal response to source location. A sensitive trans-impedance photodiode amplifier was in-house developed, enclosed in an aluminium shielded box to maintain protection from external interfering sources, connected to data logger through a coaxial cable out to the control room. A solid-water phantom, with a catheter inserted the whole way through, was CT scanned, and dwell positions were digitised using the treatment planning system (TPS). The fibres were placed within a solid-water phantom at varying distances above a 192-Ir brachytherapy source located within the catheter, with dwell positions varied every 2.5 mm.

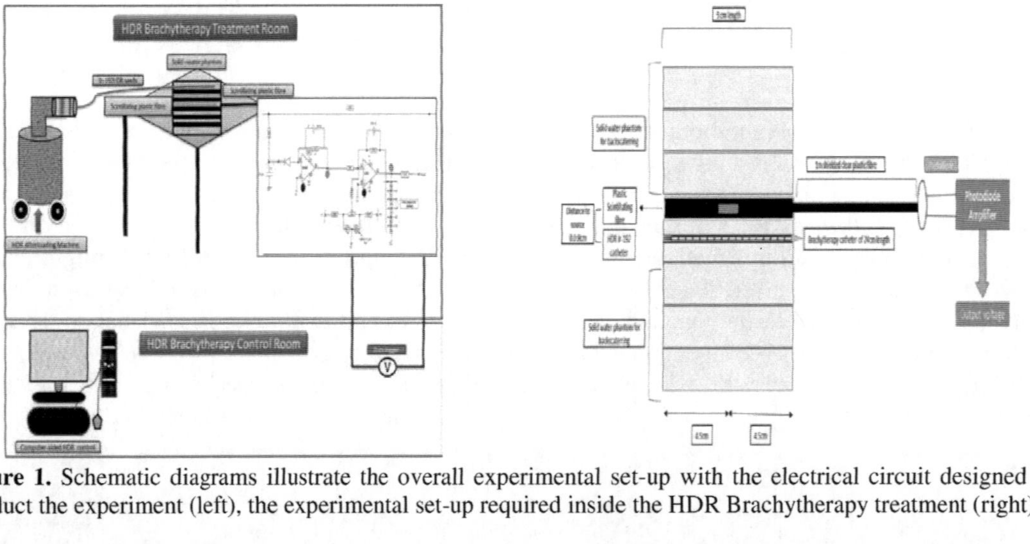

Figure 1. Schematic diagrams illustrate the overall experimental set-up with the electrical circuit designed to conduct the experiment (left), the experimental set-up required inside the HDR Brachytherapy treatment (right).

3. Results and discussion

3.1 Optical properties of the optical fibre used:

The following table, shown below, shows the main properties of the fibre probes used in this study as stated by the manufacturer (Saint-Gobain Crystals, USA). **A** denotes the clear fibre probe with no scintillating fibre attached, **B** denotes 1 m length of fully scintillating fibre, and **C** indicates the composite fibre that has 7.5 cm scintillating plastic fibre attached to 1 m clear fibre.

Table 1. Properties of the optical fibres probes.

	Fibre probe	Emission colour & Peak (nm)	Scintillation fibre attached	Length of clear fibre	Core Ref Index	Cladding Ref Index
A	Clear fibre (BCF-98)	Clear waveguide N/A	0	1 m	Polystyrene	Acrylic
B	Fully scintillating fibre (BCF-60)	Green (530nm)	1 m	0	n1= 1.60	n2=1.49
C	Partially scintillating fibre (BCF-60)	Green (530nm)	7.5 cm	92 cm		

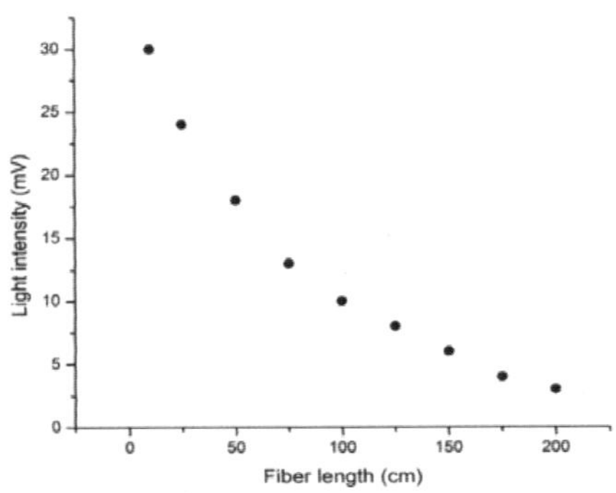

Figure 2. Attenuation of fully scintillating plastic fibre of 2 m long upon interaction with HDR 192-Ir.

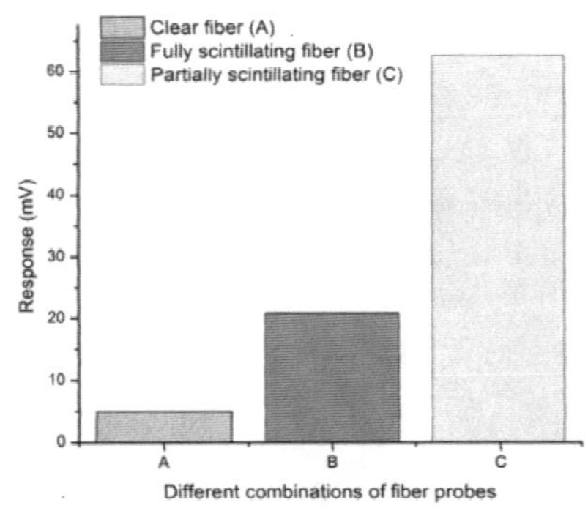

Figure 3. Comparison of three different plastic fibre probes (A, B, C) of the same length upon irradiation with HDR192-Ir, SDD = 8mm.

It is notable from Figure 1 that the fully BCF-60 green emitter, 2 m long and a diameter of 1 mm, exhibits high attenuation due to the nature of the multi-mode fibre and high probability of scintillation mechanism occurred over the whole length of the fibre probe. Attenuation would be minimal if much smaller length of BCF-60 scintillating element was optically coupled to a plastic optical fibre [22]. Hence, it is observed from Figure 2 and Figure 3 that the partially scintillating fibre (C) shows a higher signal (less attenuation) than the fully scintillating fibre (B) which demonstrates lower signal due to higher attenuation of this sort of fibre. However, the signal generated from the clear fibre (A) only being the lowest among other fibre probes and it is attributed to the effect of Cherenkov light produced at a certain threshold which is obviously above the average energy of the iridium source of 380 keV [23]. It accounts for 37.59% of the main signal for the partially scintillating plastic fibre (C) which shows constant behaviour in most conditions.

3.2 Positional sensitivity of the proposed fibre optic dosimetry system for different HDR dwell positions at different source to detector distance

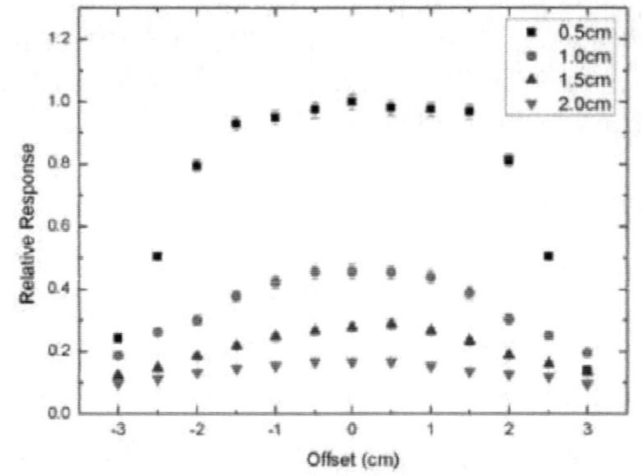

Figure 4. Monto Carlo Simulation derived responses of a 5cm length scintillating fibre with an 192-Ir source at different dwell positions within a catheter, with the catheter to detector distance of 0.5cm, 1cm, 1.5cm and 2cm.

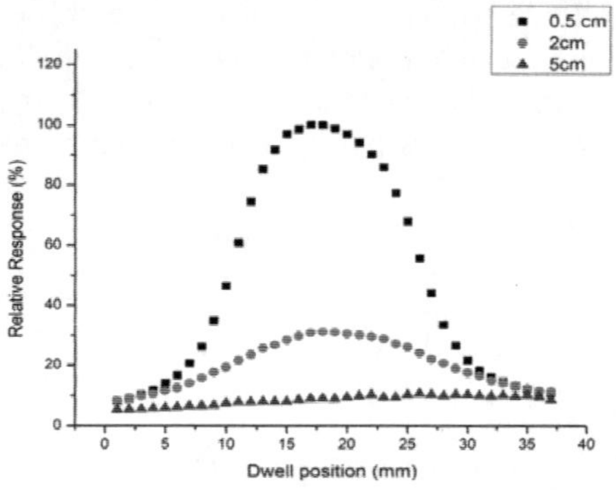

Figure 5. Response of 4cm length scintillating fibre upon irradiated with 192-Ir source at different catheter to detector distances.

The Monte Carlo simulation (Figure 4) demonstrated the variance of detector response with source location, validating the ability of a 3 cm length scintillating fibre and less to distinguish between varying source locations once signal to noise ratio (SNR) maintained high. As can be seen in Figure 5 that the initial investigation has validated the expected response of the scintillating plastic fibres for multiple dwell positions, and high resolution is notably achieved when the detector is placed close to the source (SSD=0.5 cm) and completely invisible when the detector was away from the source (SSD=5 cm)

3.3 The optimal length of scintillating fibre

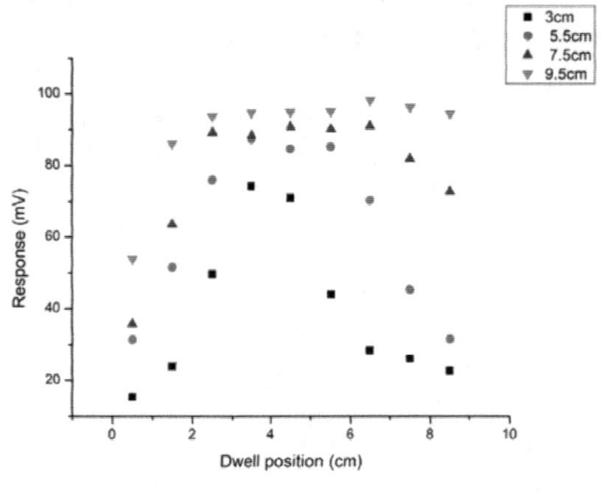

Figure 6. Positional sensitivities of different lengths of scintillating fibres, SDD = 1cm above 192-Iriduim source for multiple dwell positions.

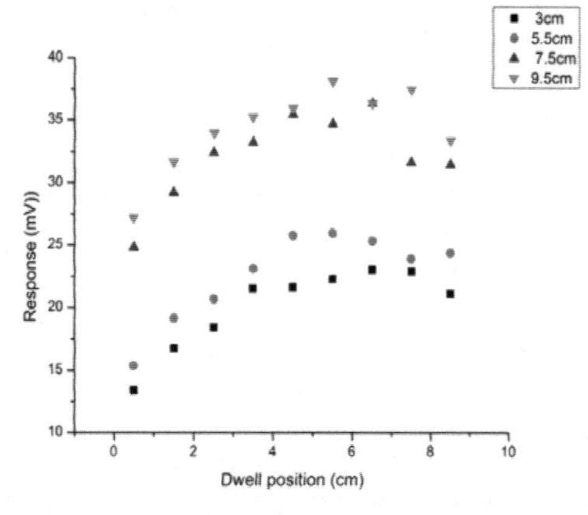

Figure 7. Positional sensitivities of different lengths of scintillating fibres, SDD = 3cm above 192-Iriduim source for multiple dwell positions.

From the above results in both Figure 6 and Figure 7, positional sensitivities were obtained for different fibre lengths and catheter to detector distances. The shorter fibre length, the less signal from the amplifier, which demonstrates poorer signal to noise (S/N) ratio. The 9.5cm of scintillating fibre has a flatted response while HDR source dwelling for a distance from 3cm to 9cm. That would be due to the large sensitive volume of the detector and the contribution of the scintillation signal along with the combination of the Cherenkov and fluorescence signal generated from the optical waveguide (the clear fibre used to transmit the scintillation signal to the photodetector). The 3cm length of scintillating fibre has clearly proven to distinguish between the responses of most adjacent dwell positions when source dwelling every 1 cm. If higher spatial resolution is to be achieved, there is a necessity to reduce the scintillation length but on the account of (S/N) ratio being relatively low. Therefore, (S/N) ratio must be kept high and higher gain must be applied to maintain reasonable signal acquisition. Positional sensitivity for each dwell position was clearly distinguishable upon placing the fibre closely to the source (SSD= 1 cm) and completely unobserved while placing the fibre away from the source (SSD= 3 cm).

4. Conclusion

The proposed fibre optic dosimetry system has been in-house designed and initially investigated. Properties of the optical fibre probes have been demonstrated in terms of attenuation and ideal fibre composition. In addition, response for all possible HDR dwell positions at different distance to the location of HDR source have been theoretically and experimentally demonstrated. The system also was characterised and optimised in terms of selecting optimal fibre length and appropriate SDDs. It proves to be eligible for further clinical investigations, such as simultaneous dose measurement and source localisation. The system has shown its capability for providing 3D position reconstruction through in-vivo source tracking.

5. Acknowledgments

Authors would like to acknowledge the assistance by Mr. Alex Pogossov for giving advice on selecting the suitable electronic components for this project. Special thanks go to the technical staff at School of Physics, CMRP for their great efforts in assisting design the photodiode amplifier. Authors also express their gratitude for colleagues at King Faisal Specialist Hospital & Research Centre (KFSHRC) for not only providing access to the Brachytherapy room but also for their remarkable efforts in establishing this collaborative research. Also, thanks go to the government of Saudi Arabia represented in Saudi Arabian Cultural Mission (SACM) for funding this project.

6. References

[1] Kertzscher G *et al* 2014 *Br. J. Radiol.* **87** 20140206
[2] Kubo H D *et al* 1998 *Med. Phys.* **25** 375-403
[3] Nath R *et al* 1997 *Med. Phys.* **24** 1557-98
[4] Hurley C *et al* 2006 *Nucl. Instrum. Meth. A* **565** 801-11
[5] Beaulieu L and Beddar S 2016 *Phys. Med. Biol.* **61** R305-43
[6] Wootton L and Beddar S 2013 *Phys. Med. Biol.* **58** 2955-67
[7] Čerenkov P 1937 *Phys. Rev.* **52** 378
[8] Beddar A *et al* 1992 *Phys. Med. Biol.* **37** 1883-900
[9] Beddar A S *et al* 1992 *Phys. Med. Biol.* **37** 1901-13
[10] Beierholm A R *et al* 2014 *Radiat. Meas.* **69** 50-6
[11] Jang K W *et al* 2009 *Opt. Rev.* **16** 383-6
[12] Moon J *et al* 2012 *Appl. Rad. Isotop.* **70** 2627-30
[13] Aoyama T *et al* 1997 *Med. Phys.* **24** 1235-9
[14] Lacroix F *et al* 2010 *Med. Phys.* **37** 4331-42
[15] Archambault L *et al* 2008 *Phys. Med. Biol.* **53** 1865-76
[16] Jang K *et al* 2012 *Nucl. Instumen. Meth. A* **695** 322-5
[17] Lessard F *et al* 2012 *Med. Phys.* **39** 5308-16
[18] Rêgo F *et al* 2011 *arXiv preprint arXiv:1109.6545*
[19] Lambert J *et al* 2006 *Phys. Med. Biol.* **51** 5505-16
[20] Suchowerska N *et al* 2007 *Radiat. Meas.* **42** 929-32
[21] Therriault-Proulx F *et al* 2011 *Med. Phys.* **38** 2542-51

[22] Archambault L *et al* 2006 *Med. Phys.* **33** 128-35
[23] Therriault-Proulx F *et al* 2011 *Med. Phys.* **38** 2176-9

Systemic effects of synchrotron radiation

J Ventura[1,2], C N Sprung[3], H B Forrester[3], J S Palazzolo[1], A Ivashkevich[3,4], A W Stevenson[5,6], C J Hall[6], A G Georgakilas[7], P N Lobachevsky[1,8] and O A Martin[1,8,9]
[1]Research Division, Peter MacCallum Cancer Centre, Melbourne, VIC, Australia
[2]University of Melbourne Department of Obstetrics & Gynaecology and Royal Women's Hospital, Melbourne, VIC, Australia
[3]Centre for Innate Immunity and Infectious Diseases, Hudson Institute of Medical Research and Monash University, Clayton, VIC, Australia
[4]Radiation Oncology, Canberra Hospital, Garran, ACT, Australia
[5]CSIRO, Clayton, VIC, Australia; [6] Australian Synchrotron, Clayton, VIC, Australia
[7]DNA Damage Laboratory, Physics Department, School of Applied Mathematical and Physical Sciences, National Technical University of Athens, Athens, Greece
[8]Sir Peter MacCallum Department of Oncology, The University of Melbourne, Melbourne, VIC, Australia
[9]Division of Radiation Oncology, Peter MacCallum Cancer Centre, Melbourne, VIC, Australia

E-mail: olga.martin@petermac.org

Abstract. A change in an organ or tissue distant from the irradiated region was termed the radiation-induced abscopal effect (RIAE). It is not known how radiation settings affect non-targeted normal tissues and therefore the risk of radiation-related adverse abscopal effects. In a recent study, we examined abscopal effects of microbeam radiotherapy (MRT) and broad beam (BB) configurations, in mice that were locally exposed to a very short pulse of a high dose-rate synchrotron beam utilizing the Imaging and Medical Beamline (IMBL) at the Australian Synchrotron. Here we summarise this study. Oxidative DNA damage was elevated in a wide variety of unirradiated normal tissues. Out-of-field duodenum showed a trend for elevated apoptotic cell death under most irradiation conditions, however, double-strand breaks (DSBs) elevated only after exposure to lower doses. These genotoxic events were accompanied by changes in concentrations of several plasma cytokines and in frequencies of macrophages, neutrophils and T-lymphocytes in duodenum. Overall, systemic radiation responses were independent of dose, time post-irradiation, and radiation modality. These findings have implications for the planning of therapeutic and diagnostic radiation treatment to reduce the risk of radiation-related adverse systemic effects.

1. Introduction

The radiation-induced bystander effect (RIBE) is now recognised as a consequence of radiotherapy [1]. Evidence has revealed that irradiated cells not only experience the direct DNA damage effects but also exchange signals with non-irradiated distant or neighbouring cells [2,3]. RIBE has been observed in a range of cell types and tissues. Cell-cell communication *in vivo* has been termed the radiation-induced abscopal effect (RIAE), which results in changes in a tissue or organ outside the irradiation field but within the same organism [4]. Bystander/abscopal cells manifest a collection of biological consequences, including mutagenesis, chromosomal aberrations, micronucleation, neoplastic transformation, cell death, augmented proliferation and differentiation [2,5,6].

Over the past decades, various traits of RIAE have come to light. Patients were reported to experience regression of tumours distant from the irradiation site [7-10]. On the other hand, following radiotherapy, RIAE in out-of-field normal tissues can contribute to development of radiation-induced secondary malignancies [1,11] and an increased incidence of inflammatory diseases [12'13]. Paediatric patients are more prone to radiation-induced carcinogenesis [14], as their bodies are still growing, and dividing cells have been shown to be more vulnerable to somatic genetic damage due to both targeted and non-targeted effects of radiation [15,16]. Besides radiation effects, there are other factors that also contribute to the development of secondary malignancies, such as genetic susceptibility and intrinsic radiosensitivity [17,18]. Therefore, contributing factors should be studied and considered in treatment planning.

There are several mechanisms that have been proposed to regulate the RIBE. These involve the transmission of signalling molecules, which include cytokines, growth factors, free radicals, reactive oxygen and nitrogen species (ROS and RNS), either via gap junctions or secretion [19-21]. These signalling molecules interact with healthy "naïve" neighbouring cells and have the capacity to cause biochemical alterations. Non-targeted effects *in vivo*, where the initial DNA damage response (DDR) at the irradiation site activates systemic responses, involve the immune system and mediators, such as cytokines.

High dose-rate synchrotron radiation is an attractive tool commonly used to study non-targeted effects of radiation due to its precise geometry and low scatter from the parallel X-ray beam [22]. Additionally, synchrotron facilities afford the chance to perform microbeam radiation therapy (MRT). The MRT beam is comprised of an array of planar separate microbeams produced from a uniform parallel X-ray beam that is passed through a collimator. MRT is a pre-clinical modality that has shown great promise for treating cancer [23,24]. Recent reports have highlighted its ability to effectively ablate tumours while also sparing normal tissues [25]. The improved therapeutic benefit provided by MRT is thought to be associated with the fact that it delivers extremely high doses to a tumour while simultaneously preserving normal tissues due to the presence of the low dose "valley" region within the MRT dose profile [26,27]. The mechanism that supports normal tissues ability to tolerate these high doses is still poorly understood. However, various explanations have been proposed, such as rapid regeneration of vasculature [25], increased tumour cell migration [28], increased stem cell proliferation and stem cell survival in normal tissues [29], immune response alterations in tumour and normal tissues [30,31] and RIBE [32,33].

It is still not clear how irradiation settings affect out-of-field normal tissues. It has been reported that increased radiation field size and irradiation modality can increase the risk of these radiation-associated side effects [14]. However, *in vitro* experiments have demonstrated that the RIBE is dose-independent [34,35]. To address this question and to determine how radiation dose and irradiated volume impact response for out-of-field normal tissues, the systemic effects of synchrotron radiation in a mouse model have been examined [36].

2. Results

We used a combination of doses and beam configurations to irradiate mice on the right hind leg (Figure 1). After 1 and 4 days post-irradiation a selection of out-of-field normal tissues and blood samples were collected. Abscopal effects can be measured using a variety of key endpoints, including ROS [37,38], DNA damage [39-41], apoptosis [42], inflammation [43], immune response [44], cytokine concentrations [45,46], oxidative stress [47], senescence [48,49] and proliferation [50]. Our findings can be summarised as follows: in out-of-field normal tissues, all radiation settings induced 1) significant and prolonged oxidative DNA damage; 2) persistent double stranded breaks (DSBs) were induced by most irradiation settings although the effect was dose-dependent and tissue specific; 3) increasing trend of apoptotic cell death; 4) increase of oxidative and inflammatory *in-situ* markers; 5) decreased proliferation and increased occurrence of senescent cells; 6) innate and adaptive immune

IOP Conf. Series: Journal of Physics: Conf. Series **1154** (2019) 012028 doi:10.1088/1742-6596/1154/1/012028

effector responses. In addition, we found alterations in plasma levels of pro-inflammatory cytokines (Figure 1).

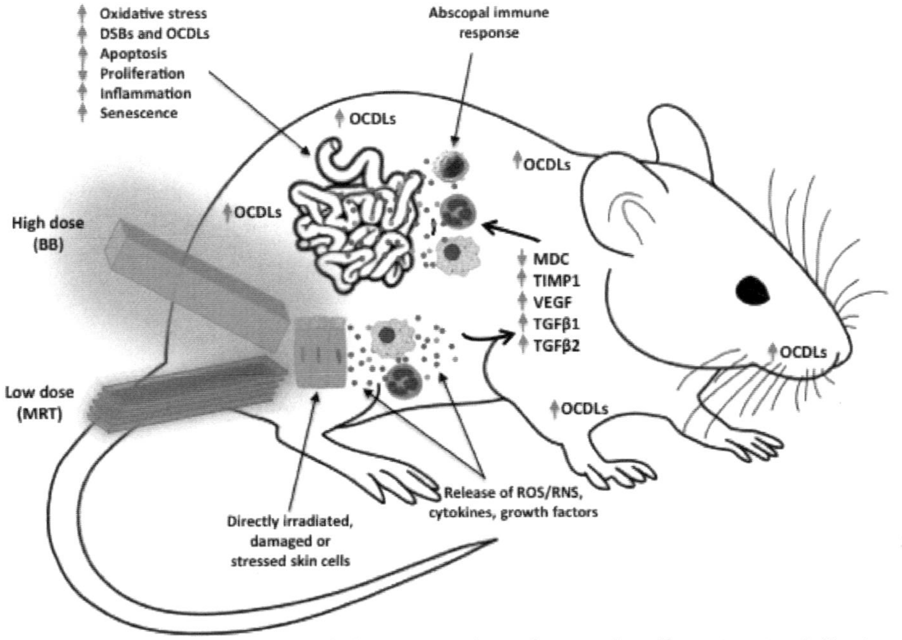

Figure 1. A schematic diagram of the propagation of systemic effects *in vivo* following localised BB and MRT synchrotron radiation. At the directly irradiated site, cells that are damaged and stressed, and immune cells are activated, releasing signalling molecules to adjacent tissues. This results to modifications to cytokine levels that can either directly or via other factors stimulate macrophages, neutrophils and T-cells in distant tissues. The imbalance that occurs in the local immune microenvironment causes the development of free radicals that consequently results to persistent oxidative stress. This leads to persistent and significant increases of OCDLs across all tissues in the body. Proliferative tissues, such as duodenum, endured significant DSBs due to unrepaired OCDLs, which ultimately encourages apoptotic cell death. The out-of-field duodenum also underwent alterations in proliferation, inflammation and senescence.

2.1 Oxidatively-induced clustered DNA lesions (OCDLs).

The OCDLs were consistently and significantly elevated across all out-of-field tissues, regardless of distance from the irradiation site, in a dose- and volume-independent manner. Bystander and abscopal effects have been documented to occur at distances up to 1 cm from the irradiated site and beyond [51-53]. We observed increases in OCDL frequencies in colon and duodenum, which are 1.5 cm from the irradiated site. Remarkably, OCDLs also increased in a tongue, which is 7 cm from the irradiation site. The lack of decline in OCDLs at 4 days post-IR was attributed to repair-resistance of OCDLs and/or persistent oxidative stress [54,55].

2.2 DSBs.

We monitored DSBs using the γ-H2AX foci assay [56]. DSBs consistently increased for out-of-field duodenum after exposure to lower but not higher peak doses. We attributed this observation to differences in γ-H2AX kinetics in directly irradiated and naïve abscopal cells. In the former, the peak γ-H2AX response is quick; the majority of radiation-induced foci disappear within hours, while in the latter, the response is slow and prolonged [51]. To further complicate this puzzle, non-targeted radiation-induced damage increases as the dose decreases; the effect is opposite in directly irradiated tissues [57,58].

2.3 Apoptosis.
Frequencies of apoptotic cells immunostained by anti-cleaved caspase 3 antibody were generally elevated in out-of-field duodenum.

2.4 Immune cells.
We established changes in frequency of macrophages/dendritic cells, neutrophils and T–lymphocytes in out-of-field duodenum, suggesting the involvement of both innate and adaptive immune responses. At the locally irradiated site (skin), macrophages/dendritic cells and neutrophils also increased, with the exception of T-lymphocytes.

2.5 Plasma cytokines.
A range of reports have provided compelling evidence that cytokines play a key role in non-targeted bystander signalling [45,46].We selected a panel of 20 cytokines associated with inflammation, DNA damage response and immune response. Blood plasma was screened for expression levels of the selected cytokines; 7 of the 20 cytokines tested were consistently and significantly altered. These include: eotaxin, macrophage derived cytokine (MDC), interleukin-10 (IL-10), TIMP metallopeptidase inhibitor 1 (TIMP-1), vascular endothelial growth factor (VEGF) and transforming growth factor beta 1 and 2 (TGFβ-1 and TGFβ-2). Some of these changes were similar to those seen in our previous study that involved plasma cytokine modulations in patients treated with fractionated radiotherapy [21,59], while some were different. The discrepancies between both studies could be attributed to differences in systemic responses between mice and humans, and between single dose and fractionated radiation treatments. In any case, alterations in cytokine expression indicate that the radiation settings induced an imbalance in the environment.

2.6 Proliferation, senescence, and oxidative and inflammatory in situ markers.
In vivo, persistent oxidative stress often leads to loss of homeostasis and to pathogenesis, such as inflammation and cancer [60]. A prime candidate that is often associated with increased oxidative stress is ROS, as they are released by directly irradiated cells or activated macrophages [61,62]. In out-of-field duodenum of BB-irradiated mice, we found increases in 8-oxoguanine a marker for oxidative stress, increases in p65, which is associated with inflammation, decreases in Ki67-positive cells, a marker for proliferation, and decreases in senescence using the Sudan Black B assay [63,64]. All of these responses were consistently and significantly altered across all BB-irradiated groups and persisted for 4 days post-irradiation.

3. Conclusions
In conclusion, we found that both MRT and BB were capable of inducing persistent systemic oxidative stress, genotoxic events and immune response, generally in a dose-independent manner. These observed bystander effects were not largely influenced by beam modality, dose and size of irradiation field. The γ-H2AX response at the higher dose differed from all other endpoints, which we believe is due to the contribution of scattered radiation confirmed by our dosimetry studies [26,40]. Our report is in agreement with the idea that macrophages and cytokines are key players in propagating the systemic effects of ionising radiation. We have confirmed that innate and adaptive immune responses influence non-targeted effects and we now aim to establish which combination of immune system components drive its propagation. This is the subject of another study, which involves the use of mice with defects in their immune system treated with similar radiation conditions described in this study. Clinical studies will validate the significance of our findings for planning of MRT and ablative radiotherapy treatment.

4. Acknowledgements

This study was mainly supported by the Australian National Health and Medical Research Council (NHMRC) grant #10275598. We thank the Australian Synchrotron for providing facilities and continuous support.

5. References

[1] Prise K M *et al* 2009 *Nat. Rev. Cancer* **9** 351-60
[2] Mothersill C *et al* 2004 *Nat. Rev. Cancer* **4** 158-64
[3] Hamada N *et al* 2007 *J. Radiat. Res.* **48** 87-95
[4] Mole R H 1953 *Brit. J. Radiol.* **26** 234-41
[5] Azzam E *et al* 2004 *Curr. Cancer Drug Tar.* **4** 53-64
[6] Hei T K 2006 *Mol. Carcinog.* **45** 455-60
[7] Kaminski J M *et al* 2005 *Cancer. Treat. Rev.* **31** 159-72
[8] Mancuso M *et al* 2012 *Curr. Mol. Med.* **12** 613-24
[9] Shiraishi K *et al* 2008 *Clin. Cancer Res.* **14** 1159-66
[10] Siva S *et al* 2015 *Cancer Lett.* **356** 82-90
[11] Hall E J *et al* 2006 *Int. J. Radiat. Oncol.* **65** 1-7
[12] Essen C F 1991 *Clin. Exp. Metastas.* **9** 77-104
[13] Ishiyama H *et al* 2012 *Clin. Genitourin. Canc.* **10** 196-8
[14] Kamran S C et al 2016 *Cancer* **122** 1809-21
[15] Sadetzki S and Mandelzweig L 2009 *Brit. J. Cancer* **100** 1021-5
[16] Dickey J S *et al* 2012 *Nucleic Acids Res.* **40** 10274-86
[17] Flint-Richter P *et al* 2007 *Lancet Oncol.* **8** 403-10
[18] Habash M *et al* 2017 *Cancer* **9** 147
[19] Zhou H *et al* 2005 *Signaling Pathway* 14641
[20] Shao C *et al* 2003 *FASEB J.* **17** 1422-7
[21] Sprung C N *et al* 2015 *Cancer Lett.* **368** 191-7
[22] Lobachevsky P *et al* 2015 Radiation Research 184 650-9
[23] Dilmanian A *et al* 2002 *J. Neuro-Oncol.* **4** 26-38
[24] Dilmanian A *et al* 2003 *Radiat. Res.* **159** 632-41
[25] Bouchet A *et al* 2010 *Int. J. Radiat. Oncol.* **78** 1503-12
[26] Curtis H J 1967 *Radiat. Res.* **7** 258-64
[27] Laissue J *et al* 2007 *Dev. Med. Child Neuro.* **49** 577-81
[28] Crosbie J C *et al* 2010 *Int. J. Radiat. Oncol.* **77** 886-94
[29] Zhong N *et al* 2003 *Radiat. Res.* **160** 133-42
[30] Sprung C N *et al* 2012 *Radiat. Res.* **178** 249-59
[31] Smilowitz H M *et al* 2006 *Neuro-Oncology* **78** 135-43
[32] Mothersill C *et al* 2014 *Dose Response* **12** 72-92
[33] Smith R W *et al* 2013 *Int. J. Radiat. Biol.* **89** 118-27
[34] Schettino G *et al* 2005 *Radiat. Res.* **163** 332-6
[35] Smilenov L B *et al* 2006 *Radiat. Prot. Dosim.* **122** 256-9
[36] Ventura J *et al* 2017 *Cancer Res.* **77** 6389-99
[37] Burdak-Rothkamm S *et al* 2007 *Oncogene* **26** 993-1002
[38] Hu B *et al* 2006 *Carcinogenesis* **27** 245-51
[39] Koturbash I *et al* 2006 *Oncogene* **25** 4267-75
[40] Lehnert B E *et al* 1997 *Cancer Res.* **57** 2164-71
[41] McGlynn P and Lloyd R G 2002 *Nat. Rev. Mol. Cell Bio.* **3** 859-70
[42] Belyakov O *et al* 2002 *Radiat. Prot. Dosim.* **99** 249-51
[43] Lorimore S A *et al* 2001 *Oncogene* **20** 7085-95
[44] Coates P J *et al* 2008 *J. Pathol.* **214** 610-6
[45] Chou C H *et al* 2007 *Clin. Cancer Res.* **13** 851-7
[46] Narayanan P *et al* 1999 *Radiat. Res.* **152** 57-63
[47] Havaki S *et al* 2015 *Cancer Lett.* **356** 43-51
[48] Gorgoulis V G and Halazonetis T D 2010 *Curr. Opin. Cell Biol.* **22** 816-27
[49] Bartek J *et al* 2008 *Nat. Cell Biol.* **10** 887-9
[50] Belyakov O V *et al* 2003 *Br. J. Cancer* **88** 767-74

[51] Sedelnikova O A *et al* 2007 *Cancer Res.* **67** 4295-302
[52] Belyakov O V *et al* 2005 *P. Natl. Acad. Sci. USA* **102** 14203-8
[53] Koturbash I *et al* 2016 *Mutat. Res.* **787** 43-53
[54] Georgakilas A G 2008 *Mol Biosyst.* **4** 30-5
[55] Gollapalle E *et al* 2007 *Radiat. Res.* **167** 207-16
[56] Bonner W M *et al* 2008 *Nat. Rev. Cancer* **8** 957-67
[57] Seymour C B and Mothersill C 2000 *Radiat. Res.* **153** 508-11
[58] Suchowerska N *et al* 2005 *Phys. Med. Biol.* **50** 3041-51
[59] Siva S *et al* 2014 *PloS one* **9** e109560
[60] Sedelnikova O A *et al* 2010 *Mutat. Res.* **704** 152-9
[61] Hussain S P *et al* 2003 *Nat. Rev. Cancer* **3** 276-85
[62] Tartier L *et al* 2007 *Cancer Res.* **67** 5872-9
[63] Georgakopoulou E A *et al* 2013 *Aging* **5** 37-50
[64] Evangelou K *et al* 2016 The Specific Histochemical Stain for Lipofuscin: A Novel Method to Detect Senescent Cells. New York, NY: Humana Press 111.

IOP Conf. Series: Journal of Physics: Conf. Series **1154** (2019) 012029 doi:10.1088/1742-6596/1154/1/012029

On-line image-guided HDR brachytherapy (IGBT) for prostate cancer using ultrasound based real-time solution

A Yeo[1, 2], K George[1], P Krishnan[1,3] and T Kron[1, 2]
[1]Department of Physical Sciences, Peter MacCallum Cancer, Melbourne, VIC, Australia
[2]School of Applied Sciences, RMIT University, Melbourne, VIC, Australia
[3]Department of Biomedical Engineering, University of Melbourne, VIC, Australia

E-mail: Adam.Yeo@petermac.org

Abstract. A new solution exists for prostate HDR brachytherapy, allowing seamless real-time procedure based on ultrasound imaging, compared a conventional approach with separate sessions for catheter insertion, CT imaging, planning and verification before treatment. This introduces a new paradigm into HDR brachytherapy procedure – no requirement for patient transfer while enabling real-time image guidance throughout the procedure including 'beam-on' time. In this work a seamless integration of real-time workflow has been designed, which demonstrates the efficacy of 'plan and treat' solution, in particular, with minimising geometric miss via a new on-line image guidance protocol.

1. Introduction

HDR brachytherapy (HDR-BT) is a mature treatment modality to precisely deliver hypofractionated radiation dose for various sites including prostate, skin, sarcoma, breast and gynaecologic cancer [1, 2]. Conformity of dose distribution is inherently superior to external beam radiotherapy (EBRT) allowing reduced toxicity for organs-at-risk adjacent to (even inside) the target. HDR-BT is particularly useful to 'boost' dose to target (e.g. prostate), combined with relatively homogeneous dose distribution administered by EBRT [3]. However, it is inevitable for conventional off-line CT-based HDR-BT to involve separate sessions with catheter insertion, simulation, planning and treatment verification, which are labour-intensive in terms of theatre scheduling as well as multidisciplinary team staffing. In addition, it is necessary to transfer patients multiple occasions throughout their treatment, which can limit dosimetric and geometric accuracy of treatment delivery, directly relating to undesirable local control and toxicity [4, 5].

Ultrasound (US) imaging system is portable and it can provide an excellent visibility of the prostate and its surrounding anatomy [5]. This fact consequently led to the developments of US-based intraoperative HDR-BT. A modern brachytherapy treatment planning system (TPS), e.g. Oncentra Prostate™ (Nucletron B.B., Veenendaal, The Netherlands), combines transrectal ultrasound (TRUS) imaging and its volumetric reconstruction techniques with conventional TPS functions such as 3D catheter reconstruction, dose calculation, optimisation and its evaluation. This solution uses the ultrasound (US) echoes of the implant needles for the definition of the needle position on two orthogonal live images, enabling TRUS image-guided 'plan and treat' procedure for prostate HDR-BT. This allows a real-time adaptive approach for any changes of anatomy at any time of the procedure. Oncentra Prostate™ has been recently implemented in Peter MacCallum Cancer Centre (PMCC) in order to administer more effective and efficient treatment procedures, compared to the

IOP Conf. Series: Journal of Physics: Conf. Series **1154** (2019) 012029 doi:10.1088/1742-6596/1154/1/012029

existing CT-based HDR solution Oncentra™ Brachy (Nucletron B.B., Veenendaal, The Netherlands). This work focuses on the clinical implementation of this new technique, including on-line HDR image-guided brachytherapy (HDR-IGBT) workflow.

2. Materials and Methods

A series of commissioning tests has been performed to implement Oncentra Prostate™ for ultrasound imaging, catheter reconstruction, dose calculation and optimisation as well as independent dose calculation system. US volumetric reconstruction and live catheter reconstruction accuracy were characterised for ProGuide (round plastic) needle and Trocar (sharp stainless steel) needles using methods described in Siebert et al.'s work [6]. Ten different depths were measured by two different observers (i.e. n = 20). An optimisation algorithm (HIPO) in Oncentra Prostate™ was compared to the existing algorithm (IPSA) in the Oncentra™ Brachy. A seamless integration of real-time workflow has been designed and demonstrated. Tolerance for online image-guidance results were set based on 20 consecutive treatment fractions.

3. Results

Motorised ultrasound 3D-imaging could accurately reconstruct volumetric images within the uncertainty defined by slice thickness. This could be confirmed by scanning both transverse and longitudinal direction with turning 'grid-on' in ultrasound unit, where the US grid was used as a digital grid phantom given the fact that grid dots are exactly equidistant (5mm) and its connected lines are perfectly parallel or orthogonal to each other – its example is depicted in Figure 2. This simple procedure could effectively detect volumetric reconstruction error within slice thickness; hence, this has been incorporated into daily QA before the real-time HDR-BT procedure. US image with modified setup provides good image contrast for target region as well as organs-at-risk with no radiation dose.

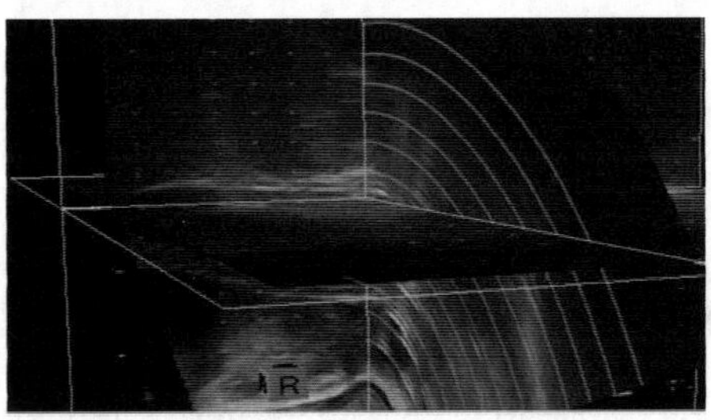

Figure 1. An example of 3D view of volumetric US image – its scan performed with US grid-on as a digital phantom to assess volumetric image reconstruction, showing negligible reconstruction error within slice thickness (0.5mm).

A 'virtual catheter' can guide catheter insertion and real-time correction of catheter reconstruction is possible based on live image. Catheter reconstruction could be accurately performed for both round plastic (ProGuide) and sharp stainless steel (Trocar) needles, based on ultrasound image within 1 mm (0.5±0.4 mm) measurable uncertainty (average error ± 2SD from 20 measurements). The trocar needle has a sharper tip providing better catheter tip definition whiel inherently more ultrasound artefact due to its high-Z component, compared to the round plastics; hence, both types of needle show equivalent catheter reconstruction accuracy.

Table 1 shows dosimetric comparison between CT-based planning with IPSA and US-based real-time planning with HIPO for 40 consecutive fractions treated in PMCC (20 patients from each technique). Compared to the former technique, the newer technique showed reduction of V150% and V200% of the target volume by 24% and 13%, respectively, while D30% and D10% of urethra are kept within 5% change, which is still far less than tolerance dose. This implies more homogeneous dose distribution for efficient planning (i.e. DVH optimisation) with an appropriate set of user input (e.g. plan objective and constraints), which agree with a previous study [7].

Table 1. Comparison of DVH parameters for target (prostate plus margin) as well as urethra between CT-based planning with IPSA and US-based real-time planning with HIPO.

	CT-based planning with IPSA	US-based real-time planning with HIPO
V_{target} (in cc)	37.3 (± 9.8) cc	36.8 (± 9.9) cc
$V_{150\%}$ (CTV)	31.9 (± 3.1) %	24.3 (± 2.4) %
$V_{200\%}$ (CTV)	12.6 (± 2.3) %	7.9 (± 1.0) %
$D_{30\%}$ (Urethra)	98.6 (± 3.1) %	104.8 (± 2.2) %
$D_{10\%}$ (Urethra)	103.1 (± 3.2) %	109.3 (± 2.8) %

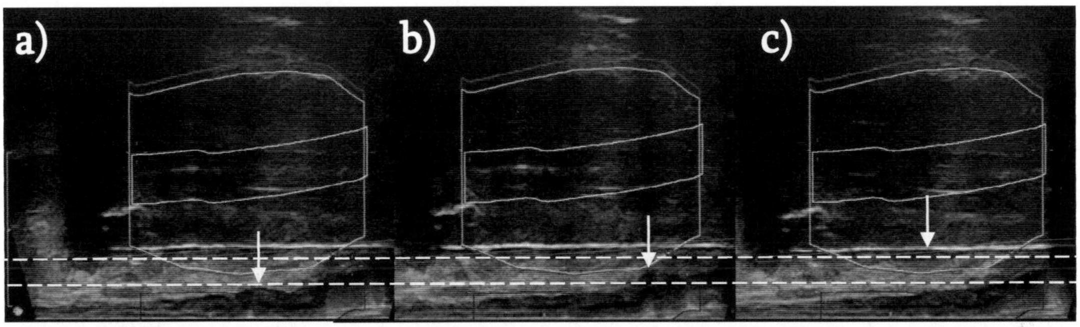

Figure 2. Three snapshots of 4D-movie file during treatment with noticeable rectal movement, causing anatomy change during beam-on. Two dashed lines and three arrows overlaid (yellow) on the images: the lower to indicate rectal movement in a) compared to the other images, and the upper to indicate live catheter shift posteriorly in c) compared to the other images c). Three arrows in each of the three snapshots highlight undesired over-dosage to rectum (i.e. closer to radiation source than what is planned).

Figure shows three snapshots of live US cine 2D-imaging during treatment, illustrating an incidence of gating events due to noticeable rectal movement, which caused entire anatomy change in the imaging field-of-view during beam-on. Two dashed lines and three arrows overlaid (yellow) on the images: the lower line to indicate rectal movement in a) and b) compared to c), and the upper to indicate live catheter shift posteriorly in c) compared to the other images. Three yellow arrows in each of the three snapshots show undesired over-dosage to rectum (i.e. closer to radiation source than what is planned). This result clearly demonstrates that 2D-live ultrasound cine image during treatment can ensure accurate treatment delivery. Based on 20 consecutive patients' treatments, action level for beam-off and thereafter immediate adaptation has been set as 3 mm for longer than 5 seconds for displacements of urethra and rectum from the planned positions, where rectum movement is only considered anteriorly towards catheters. In addition, post-treatment image acquisition and 3D reconstruction with overlaying the treated plan and structure could verify any possible deviation in patient geometry before and after the procedure dose, as like post-treatment CBCT in external beam radiotherapy, yet with nor cost neither radiation dose. This retrospective study could pick up changes between planned and actually delivered in terms of dosimetry, therefore it can possibly correlate undesired over-dosage to OARs with its consequent toxicity (if there are any).

4. Discussion

In conventional off-line CT-based HDR-BT, there is little (or no) contrast between prostate target and surrounding tissue, hence CTV contouring and PTV margin tends to be conservative [4, 5]. In contrast, US image-guided intraoperative HDR-BT provides excellent contrast between prostate and surrounding organs-at-risk, hence it seamlessly integrates all of separate sessions in the conventional technique, enabling real-time workflow. There are several advantages in the new technique; first of all, this single non-stop procedure can minimise patient movement and implant shift. If there are any

IOP Conf. Series: Journal of Physics: Conf. Series **1154** (2019) 012029 doi:10.1088/1742-6596/1154/1/012029

changes, the new technique enables on-line adaptive HDR-BT to account for geometric changes of anatomy as well as sub-optimal catheter placement at any time point throughout the procedure. In addition, this solution provides virtual planning option to visualise what patients will get before catheter insertion, which will be particularly useful for less experienced brachytherapy team. Most importantly, US-based real-time HDR-BT enables on-line image-guided treatment option to monitor anatomy changes and to actively act on it by 'beam-off' (gating). It can be argued that prostate moves together with catheters (radiation source), hence, there is no change in target coverage; however, distance from the catheters to OARs (e.g. rectum, urethra or bladder) can be changed – reduction in the distance is detrimental, therefore there can be undesirable over-dosage to surrounding tissues.

On the other hand, US-based real-time technique has its own limitations. The procedure requires at least two image acquisitions before and after catheter intersion. Accurate implant placement, planning and treatment are based on an assumption that the two images share a same image coordinate origin, (i.e. identical TRUS probe position relative to anatomy between the two images), which is represented by a transverse transducer of TRUS probe at a base-plane of prostate. It is a non-trivial to validate this assumption due to two reasons: i) anatomy changes after catheter insertion – prostate is pushed in superiorly and stretched as well as swollen, and ii) US image after implant is subject to time-of-flight artefact, caused by heterogeneity from the catheters. Hence, the two images before and after implant always look different. In addition, there are also similar artefacts at the tip of catheters, which is crucial to accurately reconstruct catheters. Catheter reconstruction accuracy is deemed equivalent between Proguide (plastic) needle and Strocar (stainless steel) needle. Both cases, uncertainty due to US artefact is up to 1mm, which results agree with other study [6]. These two factors can cause discrepancies in free-length (catheter length out-of-template) between actual measurements and what TPS thinks which can be detrimental to clinical outcome. Hence, it is highly recommended to 'remember' anatomy at the base-plane by capturing its image before image acquisition without implant. Furthermore, target contouring is difficult in the image after implant due to similar US artefacts. Hence, it is important to contour on the image before implant then correlate catheter position to the contoured target, in order to minimise difference in actual anatomy and how its planned. It is undoubtable that brachytherapy team needs an expert trained for US image and its system, which is often not a usual case in routine clinical practice. All of the aforementioned limitations need to be carefully addressed to maximally enhance advantages of this new technique.

The authors consider US live image as an invaluable tool to enable on-line image-guided HDR-BT. 2D-live image plays an important role as 'a safety belt' during treatment. There were two gating events over last 12 months according to our on-line image guidance protocol shown in the result section – one for rectal movement and the other for patient-coughing induced anatomy shift causing reduced distance from catheter to rectum as well as urethra rotation. Such cases would be detrimental in the conventional procedure without on-line image guidance. The US imaging system also offers additional advantage of post-treatment image and its 3D reconstruction (like post-treatment CBCT in EBRT) – this is to further confirm how much changes are involved before and after treatment, so as to retrospectively quantify potential difference between what is planned and treated.

It is worth noting that real-time 'plan-and-treat' procedure is deemed less time-consuming, hence more efficient compared to the CT-based conventional procedure. At the same this can be more effective with improving quality of plan and its delivery. According to our experience, after the first three-month learning curve with the new technique, 0.45 day per fraction is achievable with the same number of staff as opposed to 0.65 day per fraction for the conventional technique. In other words, 30% less staff are needed per given time to treat a same number of patients.

5. Conclusion

Ultrasound-guided real-time HDR-BT solution enables a single non-stop adaptive procedure within an operation room, minimising patient movement and implant shift [8]. More importantly, on-line IGBT could be effectively implemented for real-time treatment delivery verification and estimate the dose delivery to the insertion position with zero geometric uncertainty. There are some limitations including image artefacts and reproducibility of finding a same base-plane before and after catheter insertion, which need to be addressed. It is worth noting that total procedure time could be significantly reduced without compromising the quality of treatment procedure, aiming for 3-4 hours per procedure.

6. References

[1] Banerjee S *et al* 2017 *J. Contemp. Brachythrapy* **9** 146-50
[2] Hurley C *et al* 2006 *Nucl. Instrum. Meth. A* **565** 801-11
[3] Vargas CE *et al* 2006 *Int. J. Radiat. Oncol. Biol. Phys.* **66** 416-23
[4] Fuller DF *et al* 2008 *Int. J. Radiat. Oncol. Biol. Phys.* **70** 1588-97
[5] Kovács G *et al* 2005 *Radiother. Oncol.* **74** 137-48
[6] Sibert FA *et al* 2009 *Med. Phys.* **36** 3406-12
[7] Panettieri V *et al* 2014 *J. Appl. Clin. Med. Phys.* 2014 **15** 256-66
[8] Mather M L *et al* 2003 *Phys. Med. Biol.* **48** N269-75

IOP Conf. Series: Journal of Physics: Conf. Series **1154** (2019) 012030 doi:10.1088/1742-6596/1154/1/012030

Characterization of prompt gamma rays for *in-vivo* range verification in hadron therapy: A Geant4 simulation study

M Zarifi[1], S Guatelli[1], Y Qi[1], D Bolst[1], D Prokopovich[2] and A Rosenfeld[1]
[1]Centre for Medical Radiation Physics, University of Wollongong, Australia
[2]Australian Nuclear Science and Technology Organisation, Lucas Heights, Australia

E-mail: mz659@uowmail.edu.au

Abstract. Prompt gamma (PG) rays have been proposed for *in-vivo* beam range verification during treatment delivery. As a secondary by-product emitted almost instantaneously upon ion-nuclear interaction, PG rays offer real-time tracking of the Bragg peak (BP). However their detection is challenging since they have a broad energy spectrum with interference from neutrons and stray gamma rays. Numerous approaches have been proposed to utilise PG for *in-vivo* beam range verification. In this work, Geant4 Monte Carlo (MC) simulations have been used to study the spectral, spatial, temporal and angular distribution characteristics of PG emission and detection from hadron radiation fields of varying energy. Proton, ^{12}C and ^{4}He beams irradiated homogeneous water phantoms. These studies will provide valuable information for the development of clinically suitable and reliable PG detector systems.

1. Introduction

Hadron therapy is the use of charged particle beams to deliver targeted dose to tumours for radiation therapy treatment. Unlike conventional therapeutic photon beams, hadron beams penetrate the tissue with little scattering and deposit a maximum dose near the end of the ion range, with a sharp drop-off in dose that allows better sparing of the distal healthy tissue. The maximum dose distribution, termed the Bragg peak (BP), has the potential for conformed irradiation of the targeted region. However, precisely predicting and determining the BP position during treatment planning and delivery remains a challenge [1]. Prompt gamma (PG) rays offer a non-invasive *in-vivo* means for real-time monitoring of the ion dose delivery and range. PG rays are emitted almost instantaneously (10^{-19} – 10^{-9} sec [2]) in the decay of excited nuclei arising from ion-nuclear interactions, and so they offer real-time BP tracking with high accuracy in the beam position.

The characteristics and feasibility of PG rays for *in-vivo* beam range verification have been studied, with a good correlation between the location and intensity of PG production and dose deposition observed. Detector designs have been explored for PG imaging (PGI), including mechanical collimation systems such as pinhole or knife-edge slit cameras [2, 3] and electronic collimation such as Compton cameras [4, 5]. Other beam monitoring techniques include PG spectroscopy (PGS) [6], PG timing (PGT) [7] and PG peak integration (PGPI) [8], which rely on the physical properties of PG rays such as energy and timing. The temporal properties of PG rays can not only be utilised for improving the signal-to-noise ratio in neutron rejection techniques [9], but also provide information on the beam position as it traverses the target [7]. Although advances for *in-vivo* range verification using PG rays have been made, the technology remains clinically unavailable, as PG detection is challenging due to the wide energy spectrum and strong interference background from neutrons and other stray radiation. A recent study [8] showed a simple method of BP monitoring using the PG time-of-flight (TOF) peak mean and integral.

In this paper, we investigate the emission and detection characteristics of PG rays from water irradiated with particle beams of varying energy, with focus on the potential of employing the spectral and temporal properties of detected PG rays for un-collimated hadron beam range verification.

2. Materials and Methods

The Geant4 Monte Carlo toolkit [10] (version 10.02 patch 02) has been adopted to characterise PG emission and detection from hadron irradiation of a homogeneous water (H_2O, density of 1.0 g/cm^3) phantom. Mono-energetic pencil beams of proton and ^{4}He (62, 150, 200 and 250 MeV/u) and ^{12}C (120, 285, 385 and 490 MeV/u) were simulated. The phantom was cylinder in shape (ϕ40 cm x 40 cm), positioned with its front face surface at the simulation coordinate centre, and the beam incident along the central axis of the phantom (along the z-axis), see figure 1. Particle depth-dose curves were obtained by calculating the energy deposited in the phantom along the z-axis. The PG emission time is the interval of time from the particle incidence on the phantom to the production of the gamma.

An ideal detecting sphere (ϕ100 cm) surrounds the phantom, with its centre coinciding with the simulation coordinate centre, and registers gamma rays and neutrons that reach its surface once emitted from the phantom. The PG detection time, called *TOF*, is the interval of time from the particle incidence on the phantom to the detection of the secondary gamma/neutron at the detecting sphere.

The Geant4 physics included electromagnetic (standard_opt3 physics list) and hadronic physics (QGSP_BIC_HP for protons, neutrons and pions, Binary Ion Cascade models for ions). The Radioactive decay module of Geant4 was active. HP data libraries are adopted to model neutron interactions up to 20 MeV.

In the post-simulation data analysis, we quantify various parameters of the dose and PG emission/detection. We define the particle beam range R as the BP distal fall-off position, and the PG range R_{PG} as the PG emission yield fall-off position. TOF_{PG} is denoted as the PG TOF spectra peak fall-off. All fall-off values are taken at 80%. The angular distribution of PG rays was studied with the angle θ formed from the beam direction (z-axis) with the simulation coordinate centre. The maximum of a polynomial fitted to the distribution of θ values determines the preferential position P on the detecting sphere, whereby θ_P is the preferential polar angle. An energy threshold of 1 MeV is applied to all PG data, which eliminates low-energy photon counts that do not provide useful information for range verification.

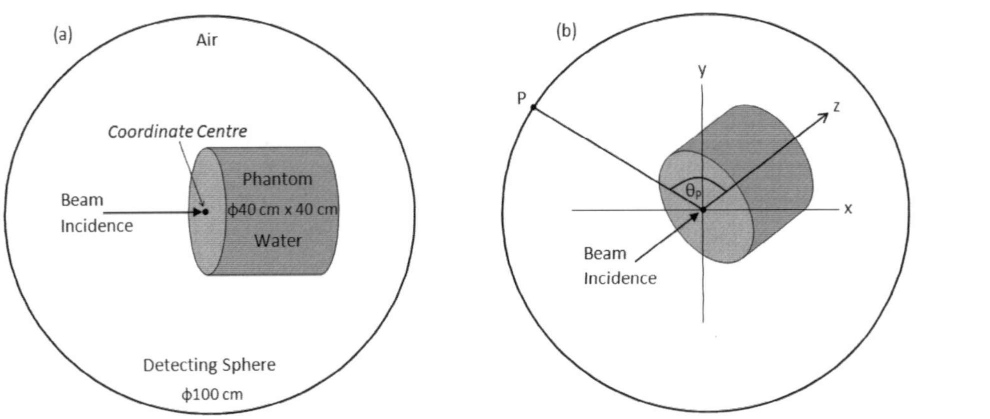

Figure 1. (a) Geometrical simulation setup. (b) The angle θ is formed from the z-axis with the coordinate centre, whereby θ_P is the preferential polar angle formed to the preferential position P on the detecting sphere determined with a polynomial fit in the post-simulation data analysis.

3. Results

For each of the particle beam types, the gamma energy spectra generated in the water phantom show characteristic PG emission peaks at 4.44, 5.21 and 6.13 MeV from the de-excitation of ^{12}C, ^{15}O and ^{16}O nuclei, respectively. A gamma peak at 511 keV represents annihilation photons, and a 2.22 MeV peak results from secondary thermal neutron capture by hydrogen. A strong spatial correlation between PG production and the dose profile was seen. The PG range accurately identifies the particle beam range: PG emission from heavier ions carbon and helium exhibit better accuracy (~2 mm) than the emission from protons (~6 mm). The time of emitted PG rays was also seen to vary with changing particle energy; with increasing beam energy PG rays are being produced at greater depths within the phantom and so generated at later times. Hence, the production time of PG rays correlates with the beam transit time and range.

The PG spatial distribution on the ideal detecting sphere showed isotropically azimuthal emission but non-isotropically axial emission such that the efficiency of PG detection can be maximised. Figure 2 compares θ_P values for PG rays from proton, ^{12}C and ^{4}He beams across the beam energies. θ_P is seen to become increasingly forward-peaked with increasing beam range, which suggests a dependence of the PG angular preferential position on the range depth with respect to the point of beam incidence. Analysis of gamma angular emission with respect to the BP position can be found in our previous study [11]. The TOF spectra of PG rays reaching the detecting sphere was also seen to depend on the particle beam range. As the beam energy increases, the TOF peak mean shifts to longer time values and the width/integral also increase, which is attributed to the greater distance of travel by the ions within the phantom, and hence the emission time of PG rays.

Figure 3 quantitatively compares the particle beam range and the corresponding PG range within the phantom, as well as PG TOF spectra data, for each particle type and energy. A good correlation is observed between R and R_{PG}. TOF_{PG} values increase gradually with increasing particle beam energy, with a slight decrease in the case for carbon beams at higher energies. Integral values of the TOF spectra exhibit a greater linearity with increasing beam energy and thus particle range. Heavier ion irradiations, such as carbon, may present a greater challenge for employing the PG TOF information for range monitoring due to the greater influence of background gamma rays such as those produced in the dose tail of the carbon beam as a result of nuclear fragmentation reactions.

Neutrons were seen to be predominantly emitted in the forward direction with longer TOF values compared to PG rays, which suggests two means (position of the detection system and timing properties of the detector) for discriminating PG rays from the interference background and hence improve the signal-to-noise ratio of PG detection.

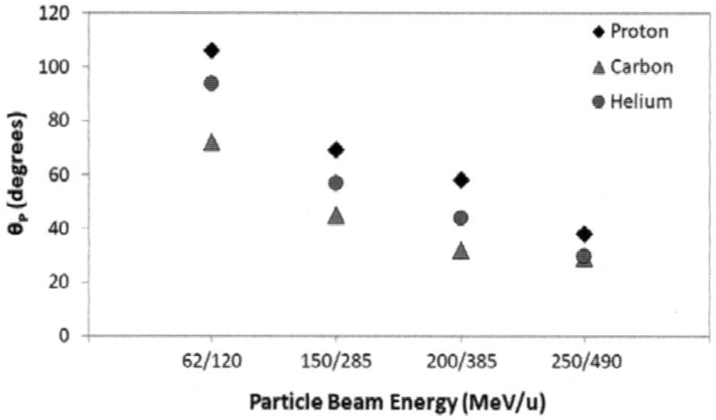

Figure 2. Preferential polar angle θ_P for PG detection, for protons and ^{4}He ions (62-250 MeV/u) and ^{12}C ions (120-490 MeV/u). An energy threshold of 1 MeV is applied to the PG data. The uncertainty of θ_P values is estimated around 1°.

IOP Conf. Series: Journal of Physics: Conf. Series **1154** (2019) 012030 doi:10.1088/1742-6596/1154/1/012030

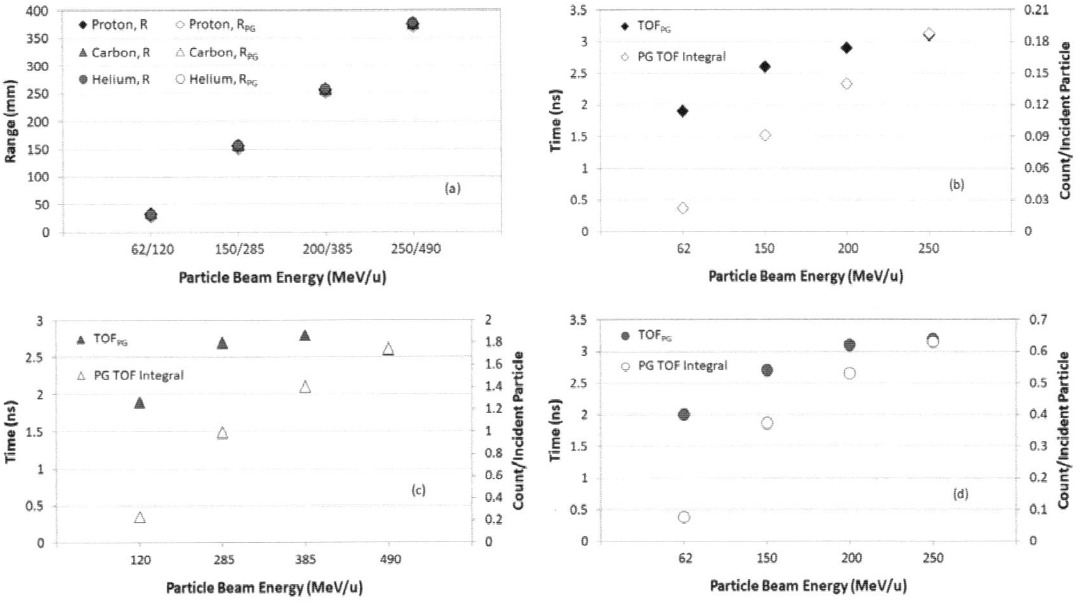

Figure 3. Quantitative data for (a) correlation between R and R_{PG}, and (b)-(d) TOF_{PG} and TOF peak integral count per incident particle, for protons, ^{12}C and ^{4}He ions, respectively. An energy threshold of 1 MeV is applied to the PG data. The uncertainty of range values is 0.1 mm and time values is 0.1 ns. The simulation statistical uncertainty affecting the count values is within 1%.

4. Conclusion

The emission and detection characteristics of PG from hadron irradiations of a water phantom were studied. Our results show that there is a preferential axial angular position for improving the efficiency of PG detection which depends on the beam energy relative to a fixed point of reference, such as the phantom surface. The range and emission time of PG rays produced within the phantom correlate to the particle beam range. As the PG emission time depends on the beam transit time, the PG TOF spectrum is also seen to change with particle range, suggesting a potentially simpler technique of beam range verification using un-collimated PG timing data. Further investigations into this technique for hadron therapy range verification are underway.

5. References

We would like to thank the University of Wollongong Information Management & Technology Services (IMTS) for computing time on the UOW High Performance Computing Cluster. This research has been conducted with the support of the Australian Government Research Training Program Scholarship

6. References

[1] Gorjiara T *et al* 2012 *Med. Phys.* **39** 7071-9
[2] Smeets J *et al* 2012 *Phys. Med. Biol.* **57** 3371-405
[3] Min C H *et al* 2006 *Appl. Phys. Lett.* **89** 183517
[4] Richard M H *et al.* 2011 *IEEE Trans. Nucl. Sci.* **58** 87-94
[5] Peterson S W *et al* 2010 *Phys. Med. Biol.* **55** 6841-56
[6] Verburg J M and Seco J 2014 *Phys. Med. Biol.* **59** 7089-106
[7] Golnik C *et al* 2014 *Phys. Med. Biol.* **59** 5399-422
[8] Krimmer J *et al* 2017 *Appl. Phys. Lett.* **110** 154102
[9] Biegun A K *et al* 2012 *Phys. Med. Biol.* **57** 6429-44
[10] Agostinelli S *et al.* 2003 *Nucl. Instrum. Methods Phys. Res. A.* **506** 250-303
[11] Zarifi M *et al* 2017 *Phys. Med.* **33** 197-206

Institute of Physics
Dirac House, Temple Back
Bristol BS1 6BE UK

ISSN: 1742-6588
ISBN 978-1-5108-8335-2
9 781510 883352

CURRAN ASSOCIATES INC.
proceedings.com

9th International Workshop on Decoherence, Information, Complexity and Entropy (DICE 2018)

Spacetime-Matter-Quantum Mechanics

Journal of Physics: Conference Series Volume 1275

Castiglioncello, Italy
17 - 21 September 2018

IOP
Institute of Physics